GEORGES CUVIER

GEORGES CUVIER

An Annotated Bibliography of His Published Works

Compiled by Jean Chandler Smith

SMITHSONIAN INSTITUTION PRESS
WASHINGTON AND LONDON

Editor and typesetter: Peter Strupp/Princeton Editorial Associates
Designer: Linda McKnight
Proof editor: Frances Kianka

Frontispiece: Georges Cuvier represented at about the age of forty. This is one of the few portraits of his middle years. Courtesy Bibliothèque Nationale.

Library of Congress Cataloging-in-Publication Data

Smith, Jean Chandler
Georges Cuvier : An annotated bibliography of his published works / Compiled by Jean Chandler Smith.
p. cm.
Includes index.
ISBN 1-56098-199-7
1. Cuvier, Georges, 1769–1832—Bibliography. 2. Natural history—Bibliography. 3. Naturalists—France—Biography. I. Title.
Z8207.3.S63 1992
[QH31.C89]
016.5914—dc20 92-7023
CIP

British Library Cataloguing-in-Publication Data is available

Manufactured in the United States of America
99 98 97 96 95 94 93 5 4 3 2 1

∞ The paper used in this publication meets the minimum requirements of the American National Standard for Permanence of Paper for Printed Library Materials Z39.48-1984

CONTENTS

Cuvier on the rostrum of the Académie des sciences. Lithograph by Albert Chireau.
Courtesy Bibliothèque Nationale.

FOREWORD

It is often said of Talleyrand that his greatest achievement, even beyond his consummate diplomacy as France's representative at the Congress of Vienna, was to serve every government from Louis XVI through the revolution, Napoleon, the restored monarchy, to Louis-Philippe—and die in bed. Georges Cuvier (1769–1832), who was, among many other things, the greatest naturalist that France has ever produced, surely runs a close second. As a young man, he served the Revolution (in a minor post as secretary to the commune of Bec-aux-Cauchois in Normandy). He later became a favorite of Napoleon, continued to serve the Bourbon restoration in equally important positions (Louis XVIII and Charles X), stayed on under Louis-Philippe after the Revolution of 1830, and also died in bed.

Cuvier often claimed that he disliked politics, but he (as a most consummate and effective politician) could only have meant that he abjured wrangling and partisan commitments (and was thereby able to serve all governments, continually accumulating influence and power throughout his life). He held a raft of positions within the educational establishment, served on the Conseil d'état for twenty years, and was, as Commissaire du roi, responsible for presenting official bills before the legislature. He also used his formidable skills in diplomacy, writing, and oratory to expand his power within science. He was the leading professor at the Muséum d'histoire naturelle, but he wielded even more influence in his thirty-year reign as perpetual secretary of the Classe des sciences physiques of the Académie des sciences. Here he presented annual reports on the status, history, and accomplishments of a wide variety of disciplines. He also wrote *éloges* for deceased colleagues. Cuvier shamelessly used this "bully pulpit" to advance his view of science under the guise of objective documentation. His éloge of Lamarck, for example, is one of the most deprecatory and chillingly partisan biographies I have ever read—though he was supposedly writing respectful comments in the old tradition of *de mortuis nil nisi bonum*.

Amidst all this, Georges Cuvier was the greatest natural historian of his, and perhaps of any, time. He is regarded as the founder of both modern comparative anatomy and vertebrate paleontology. He established the fact of extinction by careful comparison of fossil mammals with their closest living representatives. We may be tempted, quite wrongly, to view such a finding as trivial and obvious. But many scientists in Cuvier's day, including Lamarck and Thomas Jefferson (in a famous paper written before his presidency), denied the possibility of extinction and argued that apparently defunct forms would be found living in remote quarters of the earth. If extinction does not occur, then life has no directional history, we cannot use fossils to establish a geological time scale, and the earth therefore has no ascertainable history either. Thus, extinction validated history itself—and no discovery could have been more important.

Cuvier wrote hundreds of scientific papers, books, and monographs. But three monumental works stand out as powerful, comprehensive documents that established professions and set a good part of the course of nineteenth-century biology—the five-volume *Leçons d'anatomie comparée* of 1800–1805 (which presented Cuvier's adaptationist approach to studying the optimal functioning of organs and parts), the four-volume *Recherches sur les ossemens fossiles de quadrupèdes* of 1812 (which established the fact of extinction and the utility of fossils in providing a framework for geological time), and *Le règne animal* of 1817 (which codified Cuvier's celebrated classification of animals into four embranchements—vertebrates, articulates, mollusks, and radiates).

Since any of these activities would have taken a lifetime for most intellectuals, even the maximally engaged and committed, we must ask how Cuvier could possibly have accomplished so much. Our best insight into a solution may be gleaned from a famous statement of Charles Lyell, the great English geologist who opposed Cuvier's catastrophic world view but who stood in awe of his productivity. Lyell visited Cuvier in Paris and wrote:

> I got into Cuvier's sanctum sanctorum yesterday, and it is truly characteristic of the man. In every part it displays that extraordinary power of methodising which is the grand secret of the prodigious feats which he performs annually without appearing to give himself the least trouble. . . . There is first the museum of natural history opposite his house, and admirably arranged by himself, then the anatomy museum connected with his dwelling. In the latter is a library disposed in a suite of rooms, each containing works on one subject. There is one where there are all the works on ornithology, in another room all on ichthyology, in another osteology, in another *law* books! etc., etc. . . . The ordinary studio contains no bookshelves. It is a longish room, comfortably furnished, lighted from above, with eleven desks to stand to, and two low tables, like a public office for so many clerks. But all is for the one man, who multiplies himself as author, and admitting no one into his room, moves as he finds necessary, or as fancy inclines him, from one occupation to another. Each desk is furnished with a complete establishment of inkstand, pens, etc. . . . There is a separate bell to several desks. The low tables are to sit to when he is tired. The collaborators are not numerous, but always chosen well. They save him every mechanical labour, find references, etc., are rarely admitted to the study, receive orders and speak not.

We may admire Cuvier's efficiency, but recognize also that such achievement requires undiminished passion and commitment throughout life, the most precious commodity of all (and so subject to waning even among the most brilliant people).

The irony, almost the cruelty, of Cuvier's subsequent, posthumous reputation lies in our neglect and downgrading of his prodigious accomplishments. He has not been forgotten, of course, but he has been marginalized—a circumstance to which I apply the term *ironic* because no man cultivated his reputation more assiduously during life, and with an eye always turned to a later permanent renown, than Cuvier. I celebrate Jean Smith's wonderful bibliography, and I feel honored to be the writer of its foreword. But this book should not have been necessary; such a definitive work should have been published generations ago! We downplay Cuvier today for a simple reason that could not be more unjust and perverse: we regard him as having been wrong in his major scientific beliefs, particularly in two errors (one for each of the fields that he founded): his anti-evolutionism in comparative anatomy, and his catastrophism in paleontology.

I regard this neglect as grossly unfair for two basic reasons. First of all, Cuvier was not entirely wrong on either of his apostasies-by-modern-standards. He did oppose evolution (thus reflecting the dominant opinion of his time), but he did so from a perspective that would provide a foundation for Darwin's reconstruction of biology. Cuvier believed that all the parts of organisms were complexly and ideally adapted to their conditions of life. (He argued that evolution could not occur because a change in any part would demand corresponding alterations in every other organ in order to provide sustained, optimal adaptation. Since such wholesale alteration is inconceivable, evolution cannot occur. Cuvier was correct in his assertion about the impossibility of such concerted change, but wrong in his assumption that "correlation of parts"—his term for the concept—is so tight. In fact, parts can be dissociated and evolve at different rates—a phenomenon called "mosaic evolution." Darwin embraced this notion of adaptation as a centerpiece for his mechanism of natural selection. In the *Origin of Species*, he speaks of adaptation as "the expression of conditions of existence so often insisted upon by the illustrious Cuvier." As for Cuvier's catastrophism in geology, Lyell may have banished such a view to a temporary periphery of derision, but it has come back in spades through such exciting (and probably correct) ideas as the extraterrestrial impact theory of mass extinction.

Second, and more important, nothing could be more arrogant, more conducive to the misunderstanding of history, and more unfair than our tendency to judge brilliant thinkers of the past according to our current ideas of truth value for their major beliefs. Importance and truth value have no strong correlation. Mendel was ignored in his lifetime; Aristotle was mighty important, but quite wrong about the physics of gravitation. Thomas Aquinas built a system (with many wrong elements) that strongly influenced human thought for centuries, but most of us today have barely heard of him. Judge a thinker substantively by brilliance and influence, not by any retrospective criterion of a transitory current belief. Judge him or her ethically (if you must) by the primary criterion of Hippocrates: *primum non nocere* (above all, do no harm). On both scores, Cuvier should regain his place with Darwin as the greatest of natural historians.

Stephen Jay Gould
Museum of Comparative Zoology
Harvard University
Cambridge, Massachusetts

PREFACE

Georges Cuvier, 1769–1832, is known today chiefly for his contributions in the field of natural history. He is described by Stephen Jay Gould as "perhaps the finest intellect of the 19th century"[1] and ranks "with Darwin and Karl Ernst von Baer as the greatest of 19th century natural historians."[2] It is less well known today that his interests spanned all science, as well as the history of science, administration, education, and the arts, and that, although he disdained and disclaimed interest in politics, he was assuredly a deft politician, holding state positions under five French governments (the Revolution, Napoleon, and three French kings). Breadth of interest was not uncommon among the educated of his period, but Cuvier's involvement in diverse areas was not limited to interest. He was active in all fields, making substantial contributions in most.

In his several capacities, Cuvier was a prolific writer and speaker. His works, especially in science and the history of science, were published, republished, and augmented many times, as well as condensed, extracted, reported almost verbatim, and widely translated, creating confusion for cataloguers and others who consult his works.

As a naturalist, Cuvier is considered to be the founder of comparative anatomy and of paleontology. His contributions to the classification

---

[1] Gould, Stephen Jay, *Time's arrow . . .*, Cambridge, Mass.: Harvard University Press, 1987, p. 113.

[2] ———, *Hen's teeth*, New York: W. W. Norton, 1983, pp. 96–97.

of animals, both living and fossil, were immense at all levels. At the highest level, he introduced a basic and revolutionary new division of all animals into four major groups ("embranchements"); at the lowest level, he described a multitude of new species in most of the major categories of animals, his classification in most cases remaining generally valid today. He promoted the concept of extinction. He wrote the influential works of his era in comparative anatomy and paleontology, all republished and translated many times. Beyond that, he laid the foundations of modern methodology in both sciences. His research and elucidation of functional anatomical relationships shaped the biological and paleontological thinking of the time, providing the foundation for dynamic new concepts. On a philosophical level, he understood and contested the theoretical issues that his discoveries evoked in the pre-Darwinian struggle to discover and define evolution.

In the history of science, not only did Cuvier provide the raw material for modern historians of science, he himself was one of the early writers and lecturers on that subject. In fulfillment of his responsibility as a perpetual secretary of the Classe des sciences of the Académie des sciences, he wrote an annual review of the year's work of members of the Classe des sciences physiques, covering biological subjects including the applied, as well as chemistry, mineralogy, geology, and meteorology. Starting in 1829 at the Collège de France, he delivered a series of public lectures on the history of science from its beginnings into the nineteenth century. These lectures were widely attended and enthusiastically received both in France and abroad. In addition to his articles and books on the history of science, his competence in Latin and Greek allowed him to annotate editions of Pliny's *Natural History* as well as to evaluate the thinking of classical naturalists based on their original writings and include them in his writings on the history of science.

Cuvier was a successful administrator. For twenty-nine years (until his death), he was perpetual secretary of the Académie des sciences, Classe des sciences physiques, in his time the most prestigious and influential scientific institution in the world. He held state positions under a variety of French governments, starting at the age of twenty-four during the Revolution with a minor position in a Normandy commune.

He was a favorite of Napoleon and appreciated in later governments, serving as president of the Ministère de l'intérieur, member of the Conseil d'état, and Commissaire du roi (defending all bills brought by the ministry in both legislative chambers), as well as serving in many capacities in the field of education.

As an educator, Cuvier proved himself at various levels. As a professor at the Muséum national d'histoire naturelle, he delivered public lectures throughout his career. Since public lectures constituted the teaching responsibilities of the museum and were open, they attracted a wide variety of attendees. He also gave public lectures at the Lycée des arts and the Collège de France. In the popular public lecture series in history of science mentioned previously as well as others in comparative anatomy, and in his willingness and eagerness to demonstrate or instruct those who worked with him or visited his laboratory, Cuvier proved himself a first-line educator. In addition to being a practicing teacher, he held a number of prominent and influential positions dealing with various aspects of natural history, including professorial positions or as incumbent of a chair, at the Panthéon, Collège de France, and Muséum national d'histoire naturelle. He was a member of the council of the Université de France; he was also grand master of the Université de France; inspector general of education; vice rector of the Faculté des sciences, Université de Paris; president of the Commission de l'instruction publique; chancellor of the Université de Paris; grand master of the Protestant faculties of theology, Université de France, and more. He was also sent by Napoleon to organize, or reorganize as needed, the public education facilities in France and in the new French departments in Italy, the Netherlands, and Germany.

Cuvier's interest in the arts was evidenced chiefly through friendships and by his attendance at various salons as well as his own weekly soirées. He was a friend of Prosper Mérimée and Stendhal, who were frequent visitors to his salon. He was acquainted with Balzac, who admired his scientific mind. He was a member of the Académie française and the Académie des inscriptions et belles lettres.

In his personal life, he was less fortunate. Although very happily married, he lost all of his four children, only one, his beloved daughter Clementine, reaching adulthood.

A prodigious memory; the capability and drive to be a tireless worker; skills as an orator, educator, writer, and linguist; substantial talent as a scientific illustrator; and outstanding talent as an organizer and classifier, both in the biological sense and in administration, made the volume and effectiveness of his contributions possible.

There is no comprehensive bibliography of Cuvier's writings as there has been for other great naturalists of the eighteenth and nineteenth centuries, such as Linnaeus, Buffon, and Darwin. The present bibliography represents an attempt to remedy this neglect, and will perhaps bear witness to his activity in nonscientific undertakings.

ACKNOWLEDGMENTS

Compilers of bibliographies are probably not the most welcome of users in large libraries, especially those libraries with closed stacks: they require a large supply of books and keep them a relatively short time before returning them for reshelving. Yet in my use of national and large university libraries, I was always met with courteous, friendly, and efficient service and with no intimation that I was a burden. This reception was especially remarkable in foreign libraries, where staffs were often smaller than those in comparable United States libraries. Reference librarians responded to my information needs with the same helpfulness, even when in some cases there was a long line behind me waiting in turn for assistance. Unfortunately I cannot name individually all those who provided these services. The list of libraries in which I spent the most time and to which this appreciation is especially directed follows:

Bibliothèque Nationale, Paris
Muséum National d'Histoire Naturelle, Bibliothèque centrale, Paris
Institut de France, Bibliothèque, Paris
British Library, London
British Museum (Natural History) Department of Library Services, London
Staatsbibliothek Preussischer Kulturbesitz, Berlin
Academy of Natural Sciences, Philadelphia
Library of Congress, Washington, D.C.
National Library of Medicine, Bethesda, Maryland

I also want to thank those many individuals in Europe who responded to my mailed questionnaire for the considerable time they gave in supplying me with very helpful responses.

The library on which the greatest burden was placed—a burden that extended to a large proportion of the total staff—was my "home" library, the Smithsonian Institution Libraries. They housed me and made available their outstanding collections in the literature of natural history, and always responded generously to my frequent requests for advice and guidance. I am grateful to Dr. Robert Maloy and Dr. Barbara Smith, Directors, and to Mrs. Vija Karklins, Acting Director and Associate Director of the Smithsonian Institution Libraries during the period I was working on this project, for their help in placing the resources of the Libraries at my disposal and for their encouragement. Maureen Canick and Martin Smith, Chiefs of Central Reference and Loan Services during this period, provided space and support. Jack Marquardt, formerly Chief of Access Services, provided invaluable reference assistance. Carol Hahn and Sha-Nice Stokes, with their remarkable sleuthing talents and extensive knowledge of interlibrary loans, have tracked down many an elusive or scarce reference. I am indebted to Ellen Wells, Head, Special Collections Department, and Janet Stanley, Chief of the National Museum of African Art branch library, both of them compilers of many bibliographies, for sharing their bibliographic expertise with me and giving me general advice. I am also indebted to Silvio Bedini, former Deputy Director of the National Museum of American History and now Research Associate in the Smithsonian Institution Libraries, for his wise and kindly support and encouragement, and for his invaluable advice throughout this project.

Three individuals who have written on Cuvier have been of incalculable help through their publications and in person: the late Dr. William Coleman, Dr. Toby Appel, and Dr. Dorinda Outram. Dr. Appel suffered unduly because, at the time I was working on this project, she lived in Washington and was therefore virtually a neighbor, and as a result became the patient recipient of a variety of questions and the provider of much information. She read and advised on the introductory text as well.

Dr. Outram, in her publications and in person, provided me with Cuvier citations, many of which I would not have discovered otherwise. She herself has published specialized bibliographies of Cuvier, one of which complements this bibliography. Cuvier's éloges are not included in the bibliography because her publication[1] covered so thoroughly that segment of his writing. Her calendar of his letters[2] was a source for many of the records in the Letters section of this bibliography.

I am also deeply indebted to Mme. Catherine Hustache, formerly of the Muséum National d'Histoire Naturelle, Bibliothèque centrale in Paris, whose extensive knowledge of the history of French science of that period, and of the pertinent resources of the library, was of immense help. I am especially grateful to Mr. Anthony Harvey, former Librarian of the Department of Library Services of the British Museum (Natural History), for extending to me the hospitality of his library and its extensive collections, and also for, unasked, providing introductions to the librarians of other relevant libraries in London, including the British Library. I am indebted to Ms. Judith Catlin, who read the Preface and who translated most of the Polish and Russian titles. I am thankful that the personal computer appeared just in time, but I would never have enjoyed what it had to offer if Mr. Gregory Blair had not performed some of the mysterious rituals necessary to get the hardware and the software going, and if he had not always cheerfully rescued me from crashes and my ignorance.

Various scientists in the National Museum of Natural History, Smithsonian Institution, have graciously provided scientific assistance. Dr. Clayton Ray has been especially helpful, reading the introductory material and samples from the bibliography, and offering counsel and very constructive suggestions.

---

[1] Outram, Dorinda, "The language of natural power: the 'Éloges' of Georges Cuvier and the public language of nineteenth century science," *History of science*, 1978, 16:153–178.

[2] ———, *The letters of Georges Cuvier: a summary calendar of manuscript and printed materials . . .*, Chalfont St. Giles, England: British Society for the History of Science, 1980 (BSHS monographs, no. 2).

I am grateful to former secretary S. Dillon Ripley for aid from the Secretary's Fluid Research Fund of the Smithsonian Institution for assistance in my work abroad.

The inadequacies and errors in this bibliography persist in spite of all those who have helped me and therefore are all the more regretted. I am entirely and solely responsible.

INTRODUCTION

THE BIBLIOGRAPHY

Purpose

This bibliography was created with the objective of helping several audiences: naturalists, historians of science, and librarians. These categories of users have diverse, as well as overlapping, requirements. Naturalists' interests often involve the date of first publication of a paper and the dates of subsequent republication; they have, as well, an active interest in the history of their profession. Historians of science are interested in the man and his impact on his times. Librarians must identify and unscramble the many editions and republications of Cuvier's monographic works.

Scope

This bibliography is restricted to the published works of Cuvier, including monographs, journal and newspaper articles, lectures, speeches, various ephemera, and letters, both scientific and nonscientific. *The bibliography does not include Cuvier's "éloges,"* eulogies delivered on the occasion of the death of a prominent person of his day, nor his other biographical pieces. A register of these has already been published: Outram, Dorinda, "The language of natural power: the "Éloges" of Georges Cuvier and the public language of nineteenth century science," *History of science*, 1978, 16:153–178.

This bibliography is not exhaustive for the twentieth century, and no attempt has been made to include all the many facsimile reproductions and micropublications now appearing.

Sources

The bibliography has been compiled from searches of indexes, bibliographies, catalogues including those of the holdings of major libraries, bibliographies included in works on Cuvier, and electronic databases. Searches were not made of official archives. In addition, fourteen libraries were visited in the United States and seventeen in Europe. Questionnaires were sent to seventy-five foreign libraries, from which fifty replies were received. The bibliographies included in recent works on Cuvier, especially those of Toby Appel,[1] Dorinda Outram,[2] and the late William Coleman,[3] have been exceptionally helpful.

Every effort was made to see each work listed, preferably in its original form, otherwise in facsimile printing or photographic reproduction. If that was not possible, the fact is so indicated in the Note.

General Explanations

No nonstandard abbreviations have been used, except "pr." for printer of monographic publications.

When referred to in the Note, the term *Académie des sciences* has been used for that institution throughout its numerous name changes during the period covered: Institut de France, Classe des sciences mathématiques et physiques; Institut national; Académie royale des sciences de l'Institut de France; Académie royale des sciences; Académie des sciences, Classe des sciences mathématiques et physiques.

---

[1] Appel, Toby A., *The Cuvier-Geoffroy debate: French biology in the decades before Darwin*, New York: Oxford University Press, 1987.

[2] Outram, Dorinda, *Georges Cuvier: Vocation, science and authority in post-revolutionary France*, Manchester, England: Manchester University Press, [1984].

[3] Coleman, William, *Georges Cuvier Zoologist: A study in the history of evolution theory*, Cambridge, Mass.: Harvard University Press, 1964.

In the titles of the works listed, the professional or official titles following an individual's name have been omitted and indicated by ". . . ."

Numbers of plates are usually taken from a statement in the work itself or from standard bibliographic works. When the total was not found, plates in important heavily illustrated works were counted by the compiler and so indicated. When counted by the compiler, folded plates were counted as one. Tables were generally not indicated unless they were plates. When plates in monographs were counted, the location of the copy counted is given in the Note.

The printed catalogues of the following national libraries are cited frequently in the Note in shortened form. The list below gives the complete citation for each:

Bibliothèque Nationale. Catalogue général des livres imprimés. . . .

British Museum. Department of printed books. General catalogue of printed books. (The library itself has now become the British Library.)

British Museum (Natural History). Library. Catalogue of the books, manuscripts and drawings in the British Museum (Natural History).

JOURNALS

This section includes journal articles, newspaper articles, reports, and ephemera such as separates, pamphlets, or similar material. Some journals are also included in the section on Cuvier's annual review of science and in the section on his published letters. Reports written in the third person, such as reports of meetings, are included if it appears that they are chiefly in Cuvier's words. Very short reports (a small paragraph of one or two sentences) in newspapers or journals are not included unless the information they contain does not appear to have been published elsewhere.

Cuvier's speeches and discourses to the two legislative chambers are taken only from *Le Moniteur universel* and are included only if they are of substantial size.

Where possible, texts of two or more publications of one work have been compared and the relationships identified in the Note. If the texts were not compared, the Note so indicates.

Where text of versions differs (such as extracts, condensations, and translations), separate records are made and relationships are identified in the Note. In a few cases, which are so recognized in the Note, a slightly different text may be included in the same record as the primary article. Same-text republications are included in the same record under what appears to be the earliest printing and are listed in the Note, designated as: "Also published in. . . ."

Information given in the title about edition or illustrations is not repeated in the Note unless there is some discrepancy with the actual material. The correct edition is then given in the Note.

Titles for journal articles omit "par Cuvier" except for titles not in French. Statements about illustrations or tables appearing in the title are omitted unless there are discrepancies. Except for these omissions, titles of articles are given exactly as printed, including the punctuation and capitalization.

Journal issue numbers are not given unless they are helpful for locating articles.

Place of publication following the journal title is not given for those published in Paris.

ANALYSES DES TRAVAUX DE LA CLASSE DES SCIENCES MATHÉMATIQUES ET PHYSIQUES

This section consists of Cuvier's annual review of the year's work in science (excluding physics, mathematics, and astronomy) by members of the Académie des sciences. These reviews were published chiefly in journals and newspapers and are treated as explained in the Journals section.

For the annual review, Cuvier often quoted from his own articles, especially those that were earlier reports made to the Académie on the work of the individual being discussed in the year's work. These have not been exhaustively searched to locate duplication. Extracts from works listed in this section that are limited to a specific subject are included in the Journals section.

MONOGRAPHS

This section contains the monographic works of Cuvier, including those of which he was a joint author; those to which he was a contributor, including those in which his work is included by another author or editor; and those that he annotated.

Works in which Cuvier is listed as one of the authors or contributors, but in which his contribution is not identifiable, are generally not included. Works to which Cuvier's sole contribution was the use by the author or editor of his classification scheme are generally not included. Later editions and issues of works of which Cuvier was not the primary author are generally not included.

Separate records have been made for all editions of monographic works by Cuvier. Different issues (different printings apparently run from the same plates) are given as separate records if there are two or more differences in the title page, such as date and publisher or "edition" number; otherwise they are mentioned in the Note of the first issue, if possible to identify.

Titles of monographs are given in their entirety for all works by Cuvier, with only professional titles following his name or the names of others omitted. Works of which Cuvier was not the major (primary) author may have shortened titles.

Printers' names, designated by "pr." at the beginning of the Note, are given when found in the publication, with the geographic location of the printer if given, and if other than Paris.

ENCYCLOPEDIA ARTICLES

This section consists of Cuvier's contributions to the *Dictionnaire des sciences naturelles* and the *Dictionnaire des sciences médicales*.

LETTERS

This section includes Cuvier's published letters to various recipients. Many of the citations here were originally found in Dorinda Outram's *The letters of Georges Cuvier: a summary calendar of manuscript and printed materials . . .*, Chalfont St. Giles, England: British Society for the History of Science, 1980 (BSHS Monographs, no. 2), including some not seen by the compiler. Dr. Outram's work, which also contains citations and locations of manuscripts of letters, is valuable for its inclusiveness as well as for its annotations describing the contents of letters, both published and manuscript.

ARRANGEMENT

Journals

Journals are listed chronologically by year of publication, and within the same year, alphabetically by journal title, and within the same journal title, by volume number or date. *Translations* are listed alphabetically by language under year of publication and follow all articles in French for that year.

The *Procès verbaux* of the Académie des sciences were published for the first time in the early twentieth century (1910–1921 for the Cuvier years), but are arranged in this bibliography under the date of the presentation of the paper to the Académie des sciences.

Analyses des travaux . . .

Analyses des travaux . . . (Cuvier's annual reports on the year's work in natural sciences at the Académie) are arranged by the year covered, with the complete reports preceding extracts, and within that alphabetically by journal title. *Translations* follow the original work and within that, they are listed alphabetically by language.

Monographs

Monographs are listed alphabetically by title, and within the same title, chronologically by date of publication. Where Cuvier is a secondary author or a contributor, or where selections from his work have been included in another publication by an author or editor, the record appears alphabetically under the primary author following the segment of the section listing works having Cuvier as primary author. Works annotated by Cuvier appear under the primary author following the contributions.

Translations of monographs appear after all of the French editions unless the titles of the French editions change. In the latter case, translations follow the appropriate edition. Translations are arranged alphabetically by language, and within that alphabetically by date.

Encyclopedia Articles

Encyclopedia articles (contributions to the *Dictionnaire des sciences naturelles* and *Dictionnaire des sciences médicales*) are arranged alphabetically by the subject/title entry with the exception of the first general record.

Letters

Collections of letters to various recipients precede letters to individual recipients, which are listed alphabetically by recipient, and within recipient, by date of letter.

CHRONOLOGY

LOUIS XV, 1715–1774

1769	Born August 23, Montbéliard (Doubs)

LOUIS XVI, 1774–1792

1784–1788	Student, Académie Caroline, Stuttgart
1787	Received golden cross of Chevalier, allowing him to live with children of nobles in the Academy

REVOLUTIONARY PERIOD, 1789–1799

1788–1795	Tutor to Achille d'Héricy, son of the Marquis d'Héricy, Caen and Fiquainville, Normandy
1792	Publication of first scientific paper
1793–1795	Secretary to the commune of Bec-aux-Cauchois, Normandy
1795	March, arrived in Paris
1795	May, named to Commission temporaire des arts

This chronology was assembled from various sources, both French and English. No research was undertaken to identify the various titles nor to relate them to the functions or offices they represent. As a result there may be overlaps and redundancies among the positions given in the list.

1795–1802 July, assistant (suppléant) to Jean Claude Mertrud, professor of animal anatomy, Muséum national d'histoire naturelle

1795 December, replacing Mertrud, started public lectures on comparative anatomy at the Jardin des plantes, Muséum national d'histoire naturelle

1795–1832 December, 1795, member of the Institut de France, Première Classe (later Académie des sciences)

1796–1800 January, professor of natural history, École centrale du Panthéon

1796 January, started public lectures for children on natural history at the École centrale du Panthéon

1797/1798? *Tableau élémentaire de l'histoire naturelle des animaux* published

NAPOLEONIC PERIOD, 1799–1814

1799–1803 September, temporary secretary, Institut de France, Première Classe (later Académie des sciences)

1800–1832 January, chair of natural history, Collège de France

1800–1805 *Leçons d'anatomie comparée* published

1802–1803 One of six inspectors general of public education

1802–1832 October, professor of comparative anatomy, Muséum national d'histoire naturelle

1802–1803 November, sent to Marseilles to organize lycées

1803–1832 January, perpetual secretary, Académie des sciences

1803 Named to the Légion d'honneur by Napoleon

1803 Elected to the body appointing the Consistory of Paris (a Protestant organization)

1804 February, married Anne Marie (Coquet de Trazaile) Duvaucel, widow of Philippe Duvaucel

1804 Birth and death of his first child, an infant

1808 *Essai sur la géographie minéralogique des environs de Paris* published

1808–1832 Appointed by Napoleon to the Council of the Université de France

1808–1814 Inspector general of public education

1808–1809 President, Assembly of Professors, Muséum national d'histoire naturelle

1809–1832	Vice rector, Faculté des sciences, Université de Paris
1809–1810	Sent on trips to Italy (Florence, Genoa, Parma, Pisa, Turin) to reorganize higher education
1811	October, created Chevalier de l'Empire
1811	Sent to the Netherlands and southern Germany to reorganize higher education
1812	Death of his daughter Anne, four years old
1812	*Recherches sur les ossemens fossiles de quadrupèdes* published
1813	Returned to Italy to inspect the schools organized on his first trip in 1809–1810
1813	Death of his son Georges, seven years old
1813–1832?	Appointed to the Conseil d'état by Napoleon
1813–1816	Appointed maître des requêtes by Napoleon

LOUIS XVIII, 1814–1824

1814–1832	Councillor of state
1816	*Le Règne animal* published
1817	Visited England to study fossils and the dodo
1817	Vice president, Ministère de l'intérieur
1818–1832	Member, Académie française
1819–1832	President, Comité de l'intérieur, Conseil d'état
1819–1820	President, Commission de l'instruction publique, Ministère de l'intérieur
1819–1824	Grand master and acting master of the Université de France, resigning and being called back several times
1819	Created baron
1820	Became commissaire du roi
1821–1822	President, Commission de l'instruction publique
1822–1832	Grand master of the Protestant faculties of theology, Université de France

CHARLES X, 1824–1830

1824	Named "Grand Officier" of the Legion of Honor
1827	Named censor of the press, but refused to assume the position
1827–1832	Directeur des cultes non-catholiques, Ministère de l'intérieur

1828	Death of his daughter and last child, Clementine, twenty-two years old
1828	Elected president of the Société de Géographie

LOUIS PHILIPPE, 1830–1848

1830–1832	Member of the Académie des Inscriptions et des Belles Lettres
1830	Made his second visit to England
1831	October, made peer of France by Louis Philippe
1832	Died May 13, Paris

Cuvier was a member of about seventy learned societies, but only those in which he was active or held office have been listed. No foreign memberships have been included.

THE BIBLIOGRAPHY

Georges Cuvier is the author of all works for which a record has been made unless otherwise indicated. His name does not appear in the customary position at the top of each record.

JOURNALS

1. Mémoire sur les Cloportes terrestres. Par M. Cuvier. I.er Genre. I.ere. Sous-division. *Journal d'histoire naturelle. Choix de mémoires sur divers objets d'histoire naturelle;* 1792; 2(13): 18–31. 1 pl.

NOTE. Apparently Cuvier's first published scientific paper.

2. Anatomie de la Patella commune . . . *Patella vulgata* Lin. *Journal d'histoire naturelle. Choix de mémoires sur divers objets d'histoire naturelle;* 1792; 2(15): 81–95. 1 pl.

3. Observations sur quelques diptères. *Journal d'histoire naturelle. Choix de mémoires sur divers objets d'histoire naturelle;* 1792; 2(19): 253–258. 1 pl.

4. Mémoire sur la structure interne et externe, et sur les affinités des animaux auxquels on a donné le nom de Vers; lu à la société d'Histoire-Naturelle, le 21 floréal de l'an 3 [May 10, 1795]. *Décade philosophique;* 1795; 5(40): 385–396.

NOTE. Also appeared separately, n.p.: n.d.; 11p., 17cm. Continued by rec.9.

5. Description de deux espèces d'Insectes. L'Asile mante. (Planche II, Fig.3) *Asilus mantiformis, rufescens, fulcris pedum, primorum élongatis, femoribus anticis clavatis. Magasin encyclopédique;* 1795; [1. année] 1: 205–207.

NOTE. Plate was not found in the Library of Congress, Bibliothèque nationale, nor Muséum national d'histoire naturelle copies.

6. Lettre du citoyen Geoffroy, . . . , et du citoyen Cuvier, aux rédacteurs du Magasin Encyclopédique, sur le *Rhinocéros bicorne. Magasin encyclopédique;* 1795; [1. année] 1: 326–328.

7. Mémoire sur une nouvelle division des Mammifères, et sur les principes qui doivent servir de base dans cette sorte de travail, lu à la société d'Histoire naturelle, le premier floréal de l'an troisième [April 20, 1795] par les citoyens Geoffroy et Cuvier. *Magasin encyclopédique;* 1795; [1. année] 2: 164–190.

NOTE. This was also published separately, n.p.: n.d.; 26p. 20.5cm. There is some difference in the tables of the two publications.

8. Mémoire sur le Larynx inférieur des Oiseaux; Lu à la Société d'Histoire naturelle. *Magasin encyclopédique;* 1795; [1. année] 2: 330–358. 2 pl.

NOTE. Also published separately, n.p.: n.d.; 28p., 2pl. Translated into German, rec.115.

9. Second Mémoire sur l'organisation et les rapports des animaux à sang blanc, dans lequel on traite de la structure des *Mollusques* et de leur division en ordre, lu à la société d'Histoire naturelle de Paris, le 11 prairial an troisième [May 30, 1795]. *Magasin encyclopédique;* 1795; [1. année] 2: 433–449.

NOTE. Also published separately, n.p.: n.d.; 16p. 20cm. Follows rec.4.

10. Mémoire sur les rapports naturels du Tarsier, (*Didelphis macrotarsus* Gm.), lu à la société d'Histoire naturelle, le 21 messidor, an troisième [July 9, 1795], Par les citoyens Cuvier et Geoffroy. *Magasin encyclopédique;* 1795; [1. année] 3: 147–154.

NOTE. Also published separately, n.p.: n.d.; 8p., 20cm.

11. Histoire naturelle des Orang-Outangs. Par Et. Geoffroy, . . . et G. Cuvier, . . . *Magasin encyclopédique;* 1795; [1. année] 3: 451–463. 1 pl.

NOTE. Also published separately, Paris, Imprimerie du Magasin Encyclopédique, n.d.; 12p., 1pl., 19.5cm., and, with fewer footnotes, in the *Journal de physique;* 1798; 46: 185–191, 1pl.

12. Discours prononcé par le citoyen Cuvier, à l'ouverture du cours d'Anatomie comparée qu'il fait au Muséum national d'histoire naturelle, pour le citoyen Mertrud. *Magasin encyclopédique;* 1795; [1. année] 5: 145–155.

NOTE. Full course was published in several monographs, rec.712, rec.713, and rec. 714.

13. [Note sur la découverte de l'oreille interne des cétacés]. *Magasin encyclopédique;* 1795; [1. année] 6: 130.

14. Conjectures sur le sixième sens qu'on a cru remarquer dans les chauve-souris, lues à la société d'Histoire naturelle, le 17 ventôse [March 7, 1795?]. *Magasin encyclopédique;* 1795; [1. année] 6: 297–301.

15. Mémoire . . . , sur l'Anatomie du grand Limaçon, *Helix pomatia,* L. *Société philomatique de Paris. Bulletin;* 1791–1799; [Ser.I] 1 [Pt.1]: 88–89.

NOTE. Issue of Feb. 29/March/April 19, 1795.

16. Sur les espèces d'Éléphans, par les CC. Cuvier et Geoffroy. *Société philomatique de Paris. Bulletin;* 1791–1799; [Ser.I] 1 [Pt.1]: 90.

NOTE. Issue of April/July, 1795.

17. Observations sur le larynx du Couagga (*equus quagga* L.). *Société philomatique de Paris. Bulletin;* 1791–1799; [Ser.I] 1 [Pt.1]: 90–91.

NOTE. Issue of April/July, 1795.

18. Mémoire sur la circulation dans les animaux à sang blanc. *Société philomatique de Paris. Bulletin;* 1791–1799; [Ser.I] 1 [Pt.1]: 91–92.

NOTE. Issue of Aug./Dec., 1795. Read at the session of January 1, 1796 of the Académie des sciences.

19. [Rapport sur] un manuscrit ayant pour titre *Observations pour servir à la myologie des Singes . . .* présenté par le C[n] Alexandre Brongniart. *Académie des sciences. Procès verbaux;* 1910; 1(1795–1799): 28.

NOTE. Report read by Cuvier at the meeting of April 20, 1796 on a manuscript presented to him. Report signed by Cuvier and Lacépède.

20. Rapport . . . sur un Mémoire du Médecin Gaetano Sotira, au sujet du mouvement du sang dans le corps humain. *Académie des sciences. Procès verbaux;* 1910; 1(1795–1799): 90–92.

NOTE. Report presented at the meeting of August 28, 1796, signed also by Fourcroy and Sabatier, with Cuvier as reporter.

21. [Rapport sur] un mémoire lu dans son sein [Classe des sciences mathématiques et physiques de l'Institut national des sciences et des arts] par le citoyen Antoine-Nicolas Duchesne; sur l'établissement d'une nomenclature européenne d'histoire naturelle. *Magasin encyclopédique;* 1796; 2. année 1: 161–164.

NOTE. Report read by Cuvier to the Classe on March 10, 1796, and signed by Lamarck, Richard, and Cuvier. Published also, with minor changes, in the *Académie des sciences. Procès verbaux;* 1910; 1(1795–1799): 17–18.

22. Notice sur le squelette d'une très-grande espèce de quadrupède inconnue jusqu'à présent, trouvé au Paraguay, et déposé au cabinet d'Histoire naturelle de Madrid, rédigée par G. Cuvier. *Magasin encyclopédique;* 1796; 2. année 1: 303–310. 2 pl.

NOTE. Read before the Académie des sciences on April 10, 1796. Also published separately, n.p.: [Imprimerie du Magasin encyclopédique]; n.d.; 8p. 2pl. Translated into Spanish, rec.28, into German, rec.80, and into English, rec.27. Cuvier on p. 377 of rec.158 says the above was a short extract, with a poor copy of the figure of the complete skeleton, from the detailed report he had made to the Académie at their request. Rec.158 is another more complete and updated memoir on the same specimen published in 1804. For more information on Cuvier's publications on this specimen, see López Piñero, José, "Juan Bautista Bru (1740–1799) and the Description of the Genus *Megatherium*," *Journal of the history of biology;* 1988; 21: 147–162.

23. Mémoire sur les espèces d'Éléphans tant vivantes que fossiles, lu à la séance publique de l'Institut national le 15 germinal, an IV [April 4, 1796]. *Magasin encyclopédique;* 1796; 2. année 3: 440–445.

NOTE. The above is extracted from rec.35 which was read to the Académie des sciences on January 21, 1796 and subsequently selected for the above presentation at the public session. Publication of rec.35, the paper in which Cuvier first expressed his views on extinct animals, was delayed and was updated in the interim.

24. Mémoire sur l'organe de l'ouie dans les cétacés. *Société philomatique de Paris. Bulletin;* 1791–1799; [Ser.I] 1 [Pt.1]: 99.

NOTE. Issue of Dec.1795/March, 1796. Read at the March 6, 1796 meeting of the Académie des sciences.

25. Sur un nouveau genre de Mollusque. *Société philomatique de Paris. Bulletin;* 1791–1799; [Ser. I] 1 [Pt.1]: 105.

NOTE. Issue of June/July, 1796. Also published in the *Magasin encyclopédique;* 1796; 2. année 4: 416–417. Report on same specimen with somewhat different information was also published, rec.48.

26. Mémoire sur l'animal des *Lingules,* Brug. *Société philomatique de Paris. Bulletin;* 1791–1799; [Ser.I] 1 [Pt.1](Aug./Oct., 1796): 111. 1 pl.

NOTE. Also published in the *Magasin encyclopédique;* 1797; 2. année 6: 438–439.

27. Notice concerning the Skeleton of a very large Species of Quadruped, hitherto unknown, found at Paraguay, and deposited in the Cabinet of Natural History at Madrid. Drawn up by G. Cuvier. *Monthly magazine,* London; 1796; 2: 637–638. 1pl.

NOTE. Translation of rec.22 with omissions. Dr. Toby Appel supplied this reference. No other reference to it was found.

28. Noticia que da Gregorio Cuvier del Esqueleto de una especie de quadrúpedo grandísimo desconocido hasta ahora, que se halló en el Paraguay, y se conserva en el Gabinete de Historia natural de Madrid. In: Garriga, José. Descripçión de un esqueleto de un Quadrúpedo muy corpulento y raro, Que se conserva en el Real Gabinete de Historia Natural de Madrid. Publícala Don Joseph Garriga, . . . p.17–20. Madrid: Imprenta de la viuda de don Joaquin Ibarra; 1796. xvii, 20 5 pl. 37 cm.

NOTE. Translation attributed to Garriga by Cuvier of rec.22, cf. rec.158, p. 376. For more information on publications by Cuvier on this specimen, see López Piñero, José, "Juan Bautista Bru . . . and the Description of the Genus *Megatherium,*" *Journal of the history of biology;* 1988; 21: 147–163.

29. Dolomieu, Déodat, Georges Cuvier et al. Rapport . . . sur les questions discutées dans la Séance précédante et relatives aux collections d'histoire naturelle qui pourroient être légués à l'Institut National ou à autres établissemens publics. *Académie des sciences. Procès verbaux;* 1910; 1(1795–1799): 321–323.

NOTE. Presented on December 26, 1797 to the Académie des sciences. Signed by Dolomieu, Desfontaines, Fourcroy, Cuvier, Guyton, Lamarck, and Lacépède.

30. Note sur l'anatomie des ascidies. *Société philomatique de Paris. Bulletin;* 1791–1799; [Ser.I] 1 [Pt.2](1): 1.

NOTE. Issue of April, 1797.

31. Sur les différentes espèces de rhinocéros. *Société philomatique de Paris. Bulletin;* 1791–1799; [Ser.I] 1 [Pt.2](3): 17.

NOTE. Issue of June, 1797. Presented at a meeting of the Académie des sciences on Jan. 15, 1797. Also published in the *Société philomatique de Paris, Rapports généraux des travaux;* 1792–1798; 2: 103.

32. Sur les narines des cétacés. *Société philomatique de Paris. Bulletin;* 1791–1799; [Ser.I] 1 [Pt.2](55[i.e.54]): 26–29.

NOTE. Issue of July, 1797. Also published in the *Magasin encyclopédique;* 1797; 3. année 2: 299–305. A partially condensed German translation also was published, rec.51.

33. Sur les rates du marsouin. *Société philomatique de Paris. Bulletin;* 1791–1799; [Ser.I] 1 [Pt.2](6): 44.

NOTE. Issue of Aug./Sept., 1797. Also published in the *Magasin encyclopédique;* 1797; 3. année 3: 150–151.

34. Note sur une nouvelle espèce de guêpe cartonnière. *Société philomatique de Paris. Bulletin;* 1791–1799; [Ser.I] 1 [Pt.2](8): 57–58.

NOTE. Issue of Nov., 1797. Also published in the *Magasin encyclopédique;* 1797; 3. année 5: 146–148. Additional information is contained in rec.49.

35. Mémoire sur les espèces d'éléphans vivantes et fossiles, . . . Lue à l'Institut national le premier pluviôse an 4 [Jan. 21, 1796]. [Paris: Baudouin; 1798?]. 23p. 25cm.

NOTE. In his autobiography, rec.723, p.184, Cuvier says that in this paper, he first presented his views on extinct animals. Cuvier was asked to give the paper again at the public session of the Institut. For that presentation, rec.23 published in 1796, he extracted from the above. Due to delay in publication the above was published later than the abstract and was updated in the interim. On page 23, in the "Note additionnelle" Cuvier mentions other more recently identified species which had come to his attention since the paper was first read to the Académie des sciences "le premier pluviôse an 5" [January 20, 1797], but the *Académie des sciences, Procès verbaux;* 1910; 1(1795–1799): 6 gives the reading date as "le premier pluviôse, an 4" [January 21, 1796]. The above was also published in the *Académie des sciences, Mémoires;* 1799; 2: 1–22 with 5 pl. An extract appeared in 1800, rec.65.

36. . . . Compte . . . du travail du Cn. Tenon sur les dens de l'homme, du cheval et de plusieurs quadrupèdes dans differens ages. *Académie des sciences. Procès verbaux;* 1910; 1(1795–1799): 327–328.

NOTE. Presented by Cuvier at a meeting of the Académie des sciences on Jan. 11, 1798. Signed also by Lacépède and Richard.

37. Rapport . . . sur un Mémoire du C^n Baussard, intitulé *Observations sur les nains. Académie des sciences. Procès verbaux;* 1910; 1(1795–1799): 335.

NOTE. Presented by Cuvier at the meeting of Jan. 20, 1798. Signed also by Tenon.

38. [Rapport] de la Commission nommée pour examiner la pièce unique envoyé pour concourir au prix de physique. *Académie des sciences. Procès verbaux;* 1910; 1(1795–1799): 411–413.

NOTE. Presented to the Académie by Cuvier in the name of the Commission, June 24, 1798, and signed also by Hallé, Lassus, Fourcroy, and Sabatier. No reporter is indicated.

39. Rapport sur le Mémoire du C^n Noël, de Rouen, concernant la pêche du hareng. *Académie des sciences. Procès verbaux;* 1910; 1(1795–1799): 413–414.

NOTE. Presented by Cuvier at the June 24, 1798 meeting of the Académie des sciences. Signed also by Lacépède and Richard.

40. Rapport . . . sur un Mémoire du C^n Jadelot relativement à une tête humaine d'une grosseur extraordinaire. *Académie des sciences. Procès verbaux;* 1910; 1(1795–1799): 450–451.

NOTE. Presented by Cuvier at the August 18, 1798 meeting of the Académie des sciences. Also signed by Hallé, Jussieu, and Vauquelin.

41. Sur la manière dont se fait la nutrition dans les insectes. *Société philomatique de Paris. Bulletin;* 1791–1799; [Ser.I] 1 [Pt.2](10): 74–75.

NOTE. Issue of Jan., 1798. Report of a paper read at the Société d'Histoire naturelle in October, 1797. Published also in the *Journal de pharmacie;* 1798; 1(12): 128. Translated into German, rec.81. Full report was published in 1799, rec.54.

42. Nouvelles recherches sur les Coquillages bivalves. *Société philomatique de Paris. Bulletin;* 1791–1799; [Ser.I] 1 [Pt.2](11): 83.

NOTE. Issue of Feb., 1798. Presented at the Dec. 26, 1797 meeting of the Académie des sciences. Also published in the *Magasin encyclopédique;* 1798; 3. année 6: 32–33, and translated into German, rec.82.

43. Sur les organes de la voix dans les Oiseaux. *Société philomatique de Paris. Bulletin;* 1791–1799; [Ser.I] 1 [Pt.2](15): 115–116.

NOTE. Issue of May/June, 1798. Short summary report on paper read at the Académie des sciences, April 30, May 5 and 10, 1798. The full report was published in 1800, rec.66. The above was also published in the *Magasin encyclopédique,* 1798; 4. année 2: 162–165, and translated into German, rec.83.

44. Extrait d'un Mémoire sur les Ossemens fossiles de quadrupèdes. *Société philomatique de Paris. Bulletin;* 1791–1799; [Ser.I] 1 [Pt.2](18): 137–139.

NOTE. Issue of Aug./Sept., 1798. Continued by rec.74 and supplemented by rec.70. Also published in *Magasin encyclopédique;* 1798; 4. année 3: 145–150 and *Journal de physique;* 1798; 47: 315–317. Translated into English in 1798, rec.50, and in 1799, rec.59; and into German, rec.84.

45. Sur les vaisseaux sanguins des Sangsues, et sur la couleur rouge du fluide qui y est contenu. *Société philomatique de Paris. Bulletin;* 1791–1799; [Ser.I] 1 [Pt.2](19): 146–147. 1 pl.

NOTE. Issue of Sept./Oct., 1798. Also published in the *Journal de physique;* 1798; 47: 318. Two different German translations appeared in 1800, rec.79 and rec.85.

46. Sur les ossemens qui se trouvent dans le gypse de Montmartre. *Société philomatique de Paris. Bulletin;* 1791–1799; [Ser.I] 1 [Pt.2](20): 154–155.

NOTE. Issue of Oct./Nov., 1798. Also published in the *Magasin encyclopédique;* 1798; 4. année 4: 289–291. Translated into German, rec.86.

47. [L'usage et formation des méthodes en histoire naturelle]. *Société philomatique de Paris. Rapports généraux des travaux;* 1792–1797; [2]: 71–73.

NOTE. Publication date of above is 1798. Probably the paper read on July 4, 1796 before the Académie des sciences; cf. *Académie des sciences, Procès verbaux;* 1910; 1(1795–1799): 67.

48. [Description d'un mollusque de l'île de Bourbon]. *Société philomatique de Paris. Rapports généraux des travaux;* 1792–1797; 2: 105.

NOTE. Publication date of above volume is 1798. Describes same specimen as rec.25, but provides different information. Creates a new genus and species, *Phyllida borbonica.*

49. [Mémoire sur une nouvelle espèce de guêpe cartonnière]. *Société philomatique de Paris. Rapports généraux des travaux;* 1792–1797; 2: 105–106.

NOTE. Publication date of above volume is 1798. Contains additional information to rec.34.

50. Extract of a Memoir on the Fossil Bones of Quadrupeds. By Cit. Cuvier. *Monthly magazine,* London; [June/Sept., 1798]; 1: 413–417.

NOTE. Translation of rec.44. A different English translation was published in 1799, rec.59.

51. Ueber die Nasenlöcher und das Geruchsorgan der Ceraceen; aus einer vom B. Cuvier im Nationalinstitut vorgelesenen Abhandlung. Mag. Encycl. *Magazin für den neuesten Zustand der Naturkunde,* Jena; 1798; 1: 34–40.

NOTE. Partially condensed translation of rec.32.

52. Rapport . . . sur un Mémoire du C[n] Gail, concernant des recherches critiques sur les animaux connus des anciens sous les noms de *Panther* et de *Pardalis*. *Académie des sciences. Procès verbaux;* 1910; 1(1795–1799): 619–620.

NOTE. Read by Cuvier at the meeting of Sept. 7, 1799 and signed also by Lacépède and Richard.

53. Sur l'organisation de l'animal nommé Méduse; Par Cuvier. Lu à l'Institut national, en brumaire an VIII. *Journal de physique;* Nov./Dec., 1799; 49: 436–440. 1 pl.

NOTE. Read at the session of October 23, 1799. A résumé was published, rec.58, and was translated into German, rec.102.

54. Mémoire sur la manière dont se fait la nutrition dans les insectes. Lu à l'Institut national en vendémiaire an 6 [Sept./Oct.1797]. *Société d'histoire naturelle de Paris. Mémoires;* 1799; 1: 34–55. 1 pl.

NOTE. A brief report on the same paper was also published, rec.41. Above appeared also in the *Journal de physique;* 1799; 49: 331–447, with 1 pl., and was translated into German, rec.116. Cuvier considered this to be one of the "most perfect" papers he had written with nothing to take out and nothing to add since; cf. Flourens, Pierre. *Recueil des éloges historiques. . .* Première série: Paris, Garnier frères; 1856; p.185.

55. Sur les différences des cerveaux, considérés dans tous les animaux à sang rouge. *Société philomatique de Paris. Bulletin;* 1799–1801; [Ser.I] 2(27): 17–19.

NOTE. Issue of May/June, 1799. Translated twice into German, rec.87 and rec.100.

56. Sur la manière de respirer des Grenouilles. *Société philomatique de Paris. Bulletin;* 1799–1801; [Ser.I] 2(30): 42–43.

NOTE. Issue of Aug./Sept. 1799. This report does not appear in the journal index under Cuvier, but does appear under the subject, 'grenouilles', as by Cuvier, no first initial. It is signed 'C. V.', initials which Cuvier often used.

57. Observations nouvelles sur quelques Mollusques. *Société philomatique de Paris. Bulletin;* 1799–1801; [Ser.I] 2(31): 52–53.

NOTE. Issue of Sept./Oct., 1799. Also published in the *Magasin encyclopédique;* 1799; 5. année 4: 104–107. Translated into German, rec.101.

58. Mémoire sur l'organisation de quelques méduses. *Société philomatique de Paris. Bulletin;* 1799–1801; [Ser.I] 2(33): 69. 1 pl.

NOTE. Issue of Nov./Dec., 1799. Résumé of a paper read before the Académie des sciences October 23, 1799 which was published later the same year, rec.53. A German translation of the résumé appeared in 1801, rec.102.

59. An abstract of a Memoir upon the Fossil Bones of Animals. By Citizen Cuvier. *Journal of natural philosophy,* London; 1799; 2: 512–514.

NOTE. Translation of rec.44. It is not the same translation as rec.50.

60. Extrait d'un ouvrage sur les espèces de quadrupèdes dont on a trouvé les ossemens dans l'intérieur de la terre, adressé aux savans et aux amateurs des Sciences, . . . Imprimé par ordre de la classe des sciences mathématiques et physiques de l'Institut national, du 26 [i.e.21?] brumaire de l'an 9. [Paris: Baudouin, Imprimeur; 1800]. 8p. 25cm.

NOTE. Read at the November 12, 1800 meeting of the Académie des sciences. The British Library has 2 issues, with dates of [1800] and [1801]. The latter has errors from the [1800] printing corrected. The corrected version was also published in the *Magasin encyclopédique;* 1801; 7. année 1: 60–82; and the *Journal de physique;* 1801; 52: 253–267; and translated into Italian, rec.125. An extract was published in English, rec.98, and two shortened and condensed, but different, translations into German were published, rec.105 and rec.106.

61. Rapport . . . sur un Mémoire du C^n Brongniart, intitulé *Essai d'une classification naturelle des reptiles. Académie des sciences. Procès verbaux;* 1912; 2(1800–1804): 105–106.

NOTE. Report made by Lacépède and Cuvier at the February 10, 1800 meeting. No reporter is indicated.

62. Rapport . . . sur le moyen proposé, par le C^{n} Vidron pour faire entendre la musique aux sourds-muets. *Académie des sciences. Procès verbaux;* 1912; 2(1800–1804): 191–194.

NOTE. Presented to the Académie by Cuvier on July 10, 1800 and signed by Lacépède, Sabatier, Haüy, and Cuvier. No reporter is indicated. This is included because the report was represented "par l'organe de" Cuvier and his name appears last where the reporter's name is normally given.

63. Rapport . . . sur le Mémoire du C^{n} Gail, concernant les deux espèces de lièvre de Xénophon. *Académie des sciences. Procès verbaux;* 1912; 2(1800–1804): 212–213.

NOTE. Read by Cuvier in his name and that of Lacépède at the August 9, 1800 meeting of the Académie. Signed only by Cuvier.

64. Rapport . . . sur les moyens proposés par le C^{n} Charpentier Beaumont pour aider l'organe d'ouïe. *Académie des sciences. Procès verbaux;* 1912; 2(1800–1804): 219–220.

NOTE. Report by Cuvier and Charles presented on August 29, 1800. Reporter is not indicated. Charles is probably Jacques Alexandre César Charles.

65. Extrait d'un mémoire sur les espèces d'éléphans vivantes et fossiles, . . . Lu à l'Institut national le premier pluviôse an 4 [January 21, 1796] . . . *Journal de physique;* 1800; 50: 207–217. 3 pl.

NOTE. Extract of rec.35.

66. Mémoire sur les instrumens de la voix des oiseaux, . . .; Lu à l'Institut, les 11, 16 et 21 floréal an 6 [April 30, May 5, 10, 1798]. *Journal de physique;* 1800; 50: 426–451.

NOTE. A very short summary report was published in 1798, rec.43.

67. Note sur un pied d'oiseau fossile incrusté dans du gypse; Lue à l'Institut national le premier thermidor an 8 [July 20, 1800]. *Journal de physique;* 1800; 51: 128–132. 1 pl.

NOTE. A brief report on this was also published, rec.73.

68. Mémoire sur l'Ibis des anciens Égyptiens. *Journal de physique;* 1800; 51(Aug./Sept.): 184–192.

NOTE. Read at the May 1, 1800 meeting of the Académie des sciences. A résumé was also published in 1800, rec.72, as well as a slightly expanded version in 1804 with 3 different plates, rec.145. The above was included in a number of editions of the *Recherches sur les ossemens fossiles* and *Discours sur les révolutions . . . du globe,* which have not been compared with the above. An English translation appeared in 1800, rec.78.

69. Leçons d'anatomie de G. Cuvier . . . recueillies et publiées sous ses yeux; par C. Duméril . . . Tome I contenant les organes de mouvement; Tome II contenant les organes de sensation. Paris, Chez Baudouin, imprimeur de l'Institut National. *Magasin encyclopédique;* Nov./Dec., 1800; 6. année 4: 145–203.

NOTE. Summary by J. N. Hallé of the first seven lessons of the *Leçons d'anatomie comparée.* Also published separately; '*Extrait des leçons d'anatomie comparée de G. Cuvier . . . .*', n.p.: n.d.; 61p. All of the lectures, including those summarized above, were gathered and published as a complete monograph in 5 volumes, rec.712.

70. Sur les Tapirs fossiles de France. *Société philomatique de Paris. Bulletin;* 1799–1801; [Ser.I] 2(34): 73–74. 1 pl.

NOTE. Issue of Dec.1799/Jan.1800. Supplement to rec.44. Also appeared in the *Magasin encyclopédique;* 1799; 5. année 5: 257–259. Translated into German, rec.103, and into English, rec.77.

71. Sur le *Siren lacertina.* *Société philomatique de Paris. Bulletin;* 1799–1801; [Ser.I] 2(38): 106–107.

NOTE. Issue of April/May, 1800. Also published in the *Magasin encyclopédique;* 1800; 6. année 1: 237–238. A German translation appeared in 1801, rec.104.

72. Sur l'Ibis des anciens Égyptiens. *Société philomatique de Paris. Bulletin;* 1799–1801; [Ser. I] 2(39): 119. 1 fig.

NOTE. Issue of May/June, 1800. Read at the May 1, 1800 meeting of the Académie des sciences. Also published without the figure in the *Magasin encyclopédique;* 1800; 6. année 1: 527–528. Translated into German, rec.118. Expanded into an article, rec.68, and further expanded, rec.145. Republished in a number of the editions of the *Recherches sur les ossemens fossiles,* and of the *Discours sur les révolutions de la surface du globe. . .,* which were not compared with the above.

73. Sur les Ornitholithes de Montmartre. *Société philomatique de Paris. Bulletin;* 1799–1801; [Ser.I] 2(41): 129; (42): 141. 1 pl.

NOTE. Issues of July/Aug., and Aug./Sept., 1800. Read before the Académie des sciences July 20, 1800. Also published in the *Magasin encyclopédique;* 1800 6. année 3: 80–81. Published more fully in rec.67.

74. Addition à l'article des quadrupèdes fossiles de Montmartre. (voyez les nos. 18 et 34). *Société philomatique de Paris. Bulletin;* 1799–1801; [Ser.I] 2(42): 141.

NOTE. Issue of Aug./Sept., 1800. Continuation of rec.44. Articles referred to in title are rec.44 and rec.70. Published also in the *Magasin encyclopédique;* 1800; 6. année 3: 396. Translated into German, rec.119.

75. Sur une nouvelle espèce de quadrupède fossile, du genre de l'hippopotame. *Société philomatique de Paris. Bulletin;* 1799–1801; [Ser.I] 2(42): 142.

NOTE. Issue of Aug./Sept., 1800. Also published in the *Magasin encyclopédique;* 1800; 6. année 3: 396–397. Translated twice into German, rec.88 and rec.120.

76. Sur une nouvelle espèce de Crocodile fossile. . . . *Société philomatique de Paris. Bulletin;* 1799–1801; [Ser.I] 2(44): 159.

NOTE. Issue of Oct./Nov., 1800. Also published in the *Magasin encyclopédique;* 1800; 6. année 4: 409–410, and again in the same journal; 1801; 6. année 5: 383–384. Translated into German, rec.121.

77. [Fossil tapirs of France]. *Philosophical magazine,* London; Feb./May, 1800; 6: 368–369.

NOTE. Translation of rec.70.

78. Memoir on the Ibis of the antients Egyptians. By C. Cuvier. *Philosophical magazine,* London; 1800; 8(Oct.): 61–70. 1 pl.

NOTE. Translation of rec.68.

79. Ueber die Blutgefässe des Blutigels und die rothe Farbe der Flüssigkeit, die darin enthalten ist; von Cüvier. *Archiv für die Physiologie,* Halle; 1800; 4: 436–437.

NOTE. Translation of rec.45. A different German translation from that of rec.85.

80. Cüviers Nachricht von dem Scelette einer sehr grossen Art von bisher unbekannten Vierfüsser, welche in Paraguay gefunden und in das naturhistorische Kabinett nach Madrid gebracht ist. . . . *Archiv für Zoologie und Zootomie,* Braunschweig; 1800; 1(2): 208–215. 1 pl. [in issue 5].

NOTE. Translation of rec.22.

81. Ueber die Art wie bei den Insekten die Ernährung geschieht, von Cüvier. *Archiv für Zoologie und Zootomie,* Braunschweig; 1800; 1(2): 217–218.

NOTE. Translation of rec.41.

82. Neue Untersuchungen über die zweischaaligen Schaalthiere, von Cüvier. *Archiv für Zoologie und Zootomie,* Braunschweig; 1800; 1(2): 220–221.

NOTE. Translation of rec.42.

83. Ueber das Stimmwerkzeug der Vogel, von Cüvier. *Archiv für Zoologie und Zootomie,* Braunschweig; 1800; 1(2): 229–233.

NOTE. Translation of rec.43.

84. Ueber die fossilen Knochen von Vierfüssern, von Cüvier. *Archiv für Zoologie und Zootomie,* Braunschweig; 1800; 1(2): 237–241.

NOTE. Translation of rec.44 with footnote added (p.239). The footnote is signed with a 'W' and the article with an 'O'. The latter is probably Lorenz Oken as translator. Continued by rec.119 and supplemented by rec.103.

85. Ueber die Blutgefässe der Blutegel und die rothe Farbe der darinn enthaltenen Flüssigkeit, von Cüvier. *Archiv für Zoologie und Zootomie,* Braunschweig; 1800; 1(2): 242–443 [i.e.243].

NOTE. Translation of rec.45. Another German translation, rec.79, is not the same.

86. Ueber die Knochen welche sich im Gypse von Montmartre finden, von Cüvier. *Archiv für Zoologie und Zootomie,* Braunschweig; 1800; 1(2): 246–249.

NOTE. Translation of rec.46.

87. Vergleichende Anatomie des Gehirns. *Beiträge für die Zergliederungskunst;* 1800: 137–142.

NOTE. Translation of rec.55. The Deutsche Staatsbibliothek, Berlin, has the only copy of this journal located. It may also be held in London, but no response was received from several inquiries to the library reportedly holding it.

88. Eine neue Art von fossilen Säugetier aus dem Geschlechte des Hippopotamus, v. B. Cüvier. *Magazin für den neuesten Zustand der Naturkunde,* Jena; 1800; 2: 490–491.

NOTE. Translation of rec.75. Another German translation was also published, rec.120.

89. Rapport fait Par le citoyen Cuvier, au nom d'une Commission de l'Institut National, Pour examiner s'il n'y auroit point quelque amélioration à faire au règlement qui concerne les séances publiques? [Paris]: Baudouin; [1801?]. 4p. 4°.

NOTE. Report presented at the Académie des sciences, Dec. 26, 1800 and signed by Laplace, Lacépède, Cuvier; Naigeon, Dannou, Dacier; Camus, Mongez, Vincent.

90. Rapport . . . sur les dissertations présentées par le Docteur Husemann. *Académie des sciences. Procès verbaux;* 1912; 2(1800–1804): 366–368.

NOTE. Presented at the June 25, 1801 meeting of the Académie and signed also by Sabatier, Hallé, and Cuvier as reporter. It was presented in the name of the signers and Portal who is not listed as a signer.

91. Rapport sur le galvanisme, fait à l'Institut national. *Journal de physique;* 1801; 52: 318–321.

NOTE. Also published in the *Mémoires des sociétés savantes et littéraires de la république française;* 1801; 1: 132–136. Translated into English, rec.97.

92. Nouveau système des animaux sans vertèbres. *Magasin encyclopédique;* 1801; 6. année 6: 387–388.

NOTE. Review of Lamarck's work cited in title above. Extract from rec.576.

93. Sur les différences qui existent entre les crocodiles de l'ancien et du nouveau continent. *Magasin encyclopédique;* 1801; 7. année 3: 409–410.

NOTE. Résumé of paper, read on Sept. 3, 1801. Reported more briefly in rec.96. A greatly expanded version of the above was published in Germany, rec.117, in French. Translated into English, rec.99.

94. Nouvelles découvertes d'os fossiles. *Société philomatique de Paris. Bulletin;* 1811 [i.e.1801–1805]; [Ser.I] 3(51): 17.

NOTE. Issue of May/June, 1801. Translated into German, rec.122.

95. Extrait d'un mémoire sur les dents des poissons. *Société philomatique de Paris. Bulletin;* 1811 [i.e.1801–1805]; [Ser.I] 3(52): 25–26.

NOTE. Issue of June/July, 1801. Read at the June 10, 1801 meeting of the Académie des sciences. Translated into German, rec.123.

96. Des véritables différences entre les Crocodiles de l'ancien et du nouveau monde. *Société philomatique de Paris. Bulletin;* 1811 [i.e.1801–1805]; [Ser.I] 3(54): 41.

NOTE. Issue of Aug./Sept., 1801. Brief abstract of a paper read at the Sept. 3, 1801 meeting of the Académie des sciences. A more extensive résumé was also published, rec.93, and an expanded version was published in 1802, rec.117.

97. Experiments in galvanism. *Philosophical magazine,* London; [1801]; 10: 87–90.

NOTE. Translation of rec.91 with omission of final paragraph.

98. [Paper on fossil bones, read in the French National Institute]. *Philosophical magazine,* London; 1801; 10: 90–92.

NOTE. Translated extracts from rec.60.

99. On the Difference which exists between the Crocodiles of the old and the new Continent. *Philosophical magazine,* London; [1801]; 11: 277–278.

NOTE. Translation of rec.93. A greatly expanded version in French was also published, rec.117.

100. Ueber die Verschiedenheiten des Hirns bei allen rothblütigen Thieren, gelesen von Cüvier in der naturhistorischen Gesellschaft. *Archiv für Zoologie und Zootomie,* Braunschweig; 1801; 2(1): 189–193.

NOTE. Translation of rec.55.

101. Neue Bemerkungen über einige Weichthiere (Mollusca), gelesen von Cüvier in der philomatischen Gesellschaft. *Archiv für Zoologie und Zootomie,* Braunschweig; 1801; 2(1): 199–201.

NOTE. Translation of rec.57.

102. Ueber den inneren Bau der Quallen, gelesen von Cüvier im Nationalinstitute. *Archiv für Zoologie und Zootomie,* Braunschweig; 1801; 2(1): 201–203.

NOTE. Translation of rec.58.

103. Ueber die fossilen Tapirs in Frankreich, von Cüvier, gelesen in der philomatischen Gesellschaft. *Archiv für Zoologie und Zootomie,* Braunschweig; 1801; 2(1): 204–205.

NOTE. Translation of rec.70. Supplement to rec.84.

104. Ueber *Siren lacertina,* gelesen von Cüvier in der philomatischen Gesellschaft. *Archiv für Zoologie und Zootomie,* Braunschweig; 1801; 2(1): 218–219.

NOTE. Translation of rec.71.

105. Ueber die Arten der vierfüssigen Thiere wovon man fossile Knochen gefunden hat. An alle Gelehrte und Freunde der Wissenschaften gerichtet, vom Professor Cüvier Journal de phys. Germin. 9. *Magazin für den neuesten Zustand der Naturkunde,* Jena; 1801; 3: 295–314.

NOTE. Shortened and condensed translation of rec.60. Another and different German condensed translation was also published, rec.106.

106. Über die Arten der Säugethiere, von welchen man Knochen im Innern des Erde findet. Eine Aufforderung des Br. Geo. Cuvier an die Gelehrten und Freunde der Wissenschaften. *Magazin für die gesammte Mineralogie, Geognosie und mineralogische Erdbeschreibung,* Leipzig; 1801; 1(4): 449–469.

NOTE. Translation of rec.60 with some omission. Another German translation and condensation, rec.105, is not the same.

107. Rapport . . . sur un Mémoire . . . du C^{n} Geoffroy . . . , intitulé *Recherches sur les animaux du Nil connus des Grecs et sur les rapports*

de ces animaux avec le système théogonique des anciens Égyptiens. Académie des sciences. Procès verbaux; 1912; 2(1800–1804): 509–511.

NOTE. Read by Cuvier in his name and Lacépède's at the meeting of the Académie, June 2, 1802. Reporter is not indicated.

108. Rapport . . . sur des pièces d'anatomie artificielles en cire, présentées par le C^{n}. Laumonnier. *Académie des sciences. Procès verbaux;* 1912; 2(1800–1804): 550–551.

NOTE. Paper presented at the Académie, Aug. 18, 1802. Signed also by Sabatier, Tenon, and Pelletan.

109. Mémoire sur l'animal de la Lingule (*Lingula anatina* LAM.). *Paris. Muséum national d'histoire naturelle. Annales;* 1802; 1: 69–80. 1 pl.

NOTE. Translated into German, rec.133. Also published in *Mémoires pour servir à l'histoire et à l'anatomie des mollusques,* (rec.724), no.21.

110. Mémoire sur le *Bulla aperta* LINN., Bullaea de Lamarck. *Paris. Muséum national d'histoire naturelle. Annales;* 1802; 1: 156–164. 1 pl.

NOTE. Translated into German, rec.134.

111. Mémoire sur le *Clio borealis.* *Paris. Muséum national d'histoire naturelle. Annales;* 1802; 1: 242–250.

NOTE. Also published in *Mémoires pour servir à l'histoire et à l'anatomie des mollusques,* (rec.724), no.2, with minor changes. Translated into German, rec.135.

112. Mémoire sur le genre Tritonia, avec la description et l'anatomie d'une espèce nouvelle, *Tritonia hombergii.* *Paris. Muséum national d'histoire naturelle. Annales;* 1802; 1: 480–496. 2 pl.

NOTE. Also published with minor changes in *Mémoires pour servir à l'histoire et à l'anatomie des mollusques,* (rec.724), no.4. Translated into German, rec.136.

113. Sur les vers qui ont le sang rouge. *Société philomatique de Paris. Bulletin;* 1811 [i.e.1801–1805]; [Ser.I] 3(64): 121–122. 1 pl.

NOTE. Issue of June/July, 1802. Résumé of paper read before the Académie des sciences, Jan. 1, 1802. Translated into German, rec.131 and rec.330. Another report by Cuvier with the same conclusions also appeared in German, rec.124.

114. Sur les serpules. *Société philomatique de Paris. Bulletin;* [Ser.I] 1811; 3(65): 130. 1 pl.

NOTE. Issue of July/Aug., 1802. Translated into German, rec.132.

115. Ueber den untern Larynx der Vögel; von G. Cuvier. *Archiv für die Physiologie,* Halle; 1802; 5: 67–96. 2pl.

NOTE. Translation of rec.8.

116. Ueber die Ernährung der Insekten; vom Burger Cuvier. *Archiv für die Physiologie,* Halle; 1802; 5: 97–127. 1pl.

NOTE. Translation of rec.54.

117. Sur les véritables différences qui existent entre les crocodiles de l'ancien et du nouveau continent. *Archiv für Zoologie und Zootomie,* Braunschweig; 1802; 2(2): 162–177. 1 pl.

NOTE. In French. Greatly expanded version of rec.93. An English summary was also published, rec.99.

118. Ueber den Ibis der alten Aegyptier; im Nationale Institute gelesen von Cuvier. *Archiv für Zoologie und Zootomie,* Braunschweig; 1802; 3(1): 192–193.

NOTE. Translation of rec.72.

119. Zusatz zu dem Aufsatze über die fossilen Vierfüsser von Montmartre. *Archiv für Zoologie und Zootomie,* Braunschweig; 1802; 3(1): 199.

NOTE. Translation of rec.74. Continuation of rec.84.

120. Ueber eine neue Art von fossilen Vierfüssern von der Gattung des Flusspferds; gelesen von Cuvier in der philomatischen Gesellschaft. *Archiv für Zoologie und Zootomie,* Braunschweig; 1802; 3(1): 199–200.

NOTE. Translation of rec.75. Another German translation was also published, rec.88.

121. Ueber eine neue fossile Krokodillart; gelesen in der philomatischen Gesellschaft von Cuvier. *Archiv für Zoologie und Zootomie,* Braunschweig; 1802; 3(1): 211–212.

NOTE. Translation of rec.76.

122. Neue Endeckungen fossiler Knochen; gelesen in der philomatischen Gesellschaft von Cuvier. *Archiv für Zoologie und Zootomie,* Braunschweig; 1802; 3(1): 218.

NOTE. Translation of rec.94.

123. Auszug einer Abhandlung über die Zähne der Fische; gelesen im National=Institute von Cuvier. *Archiv für Zoologie und Zootomie,* Braunschweig; 1802; 3(1): 218–220.

NOTE. Translation of rec.95.

124. Ueber die Würmer, die rothes Blut haben. Von Herrn Cuvier. Auszug aus einem gefälligst vom Verfasser mitgetheilten Manuscripte. *Französische Annalen für die allgemeine Naturgeschichte, Physik, Chemie, Physiologie und ihre gemeinnützigen Anwendungen,* Hamburg; 1802; 4: 60–65.

NOTE. This is not a translation of rec.113, but apparently information on the same subject sent by Cuvier to the above journal.

125. Estratto di un' Opera sopra le specie di Quadrupedi, le cui ossa si sono trovate nelle viscere della terra, indirizzato ai Professori, e Dilettanti delle Scienze naturali dal celebre G. Cuvier, . . . Tradotto da Leopoldo Fabbroni di Firenze suo riconoscentissimo Allievo. *Memorie per i curiosi di agricoltura, e di economia rurale,* Naples; 1802; 6: 3–45.

NOTE. Translation of rec.60 with translator's preface added, p.3–15.

126. Note sur l'anatomie de quelques espèces d'Aplysies. *Le Moniteur universel;* 1803 July 1; (282): 1274.

NOTE. Presented at the October 27, 1802 meeting of the Académie des sciences. Also published in *Société philomatique de Paris, Bulletin;* 1811 [i.e.1801–1805]; [Ser.I] 3 (75): 193–194, issue of March/April, 1803; and in *Décade philosophique*, 1803; [38]: 321–323. Translated twice into German, rec.167 and rec.186.

127. Notice Sur l'établissement de la collection d'anatomie comparée du Muséum. *Le Moniteur universel.* Paris; 1803 September 7; (350): 1543–1544.

NOTE. Published also in Paris, *Muséum national d'histoire naturelle, Annales;* 1803; 2: 409–414. Translated into German, rec.174.

128. Mémoire Sur le genre *Laplysia* vulgairement nommé *Lièvre marin;* sur son anatomie, et sur quelques-unes de ses espèces. *Paris. Muséum national d'histoire naturelle. Annales;* 1803; 2: 287–314. 4pl.

NOTE. Also published in *Mémoires pour servir à l'histoire et à l'anatomie des mollusques,* (rec.724), no.9, with slight differences. Translated into German, rec.172.

129. Dissertation critique Sur les espèces d'écrevisses connues des anciens, et sur les noms qu'ils leur ont donnés. *Paris. Muséum national d'histoire naturelle. Annales;* 1803; 2: 368–384.

NOTE. Translated into German, rec.173. Also published in *Mémoires pour servir à l'histoire et à l'anatomie des mollusques,* (rec.724), no.22.

130. Sur le *Pennatula Cynomorium* (*Alcyonium epipetrum,* GMELIN), et sur les coraux en général. *Société philomatique de Paris. Bulletin;* 1811 [i.e.1801–1805]; [Ser.I] 3(78): 133.

NOTE. Issue of Aug./Sept., 1803. Translated into German, rec.187.

131. Ueber die Würmer, welche rothes Blut haben; gelesen im National=Institute, am 11ten Nivose des zehnten Jahrs [Jan. 1, 1802], von

Cuvier. *Archiv für Zoologie und Zootomie,* Braunschweig; 1803; 3(2): 238–240.

NOTE. Translation of rec.113. Another German translation was published in 1817, rec.330.

132. Ueber die Wurmröhren (serpula); gelesen im National=Institute, von Cuvier. *Archiv für Zoologie und Zootomie,* Braunschweig; 1803; 3(2): 243–244.

NOTE. Translation of rec.114.

133. Ueber das Thier, welches die *Lingula anatina* Lam. bewohnt, von G. Cuvier. *Paris. Muséum national d'histoire naturelle. Annalen des Nationalmuseums der Naturgeschichte,* Hamburg; 1803; Heft 1: 39–46. 1 pl. (Pl.vi).

NOTE. Translation of rec.109.

134. Abhandlung über die *Bulla aperta* Linn. (*Bullaea*) Lamarck, von G. Cuvier. *Paris. Muséum national d'histoire naturelle. Annalen des Nationalmuseums der Naturgeschichte,* Hamburg; 1803; Heft 2: 98–103. 1 pl. (Pl.xii).

NOTE. Translation of rec.110.

135. Abhandlung über die *Clio borealis,* von G. Cuvier. *Paris. Muséum national d'histoire naturelle. Annalen des Nationalmuseums der Naturgeschichte,* Hamburg; 1803; Heft 3: 153–159. 1 pl. (Pl.xvii).

NOTE. Translation of rec.111 with an omission.

136. Abhandlung über die Gattung *Tritonia,* nebst der Beschreibung und Zergliederung einer neuen Art, *Tritonia Hombergii,* von G. Cuvier. *Paris. Muséum national d'histoire naturelle. Annalen des Nationalmuseums der Naturgeschichte,* Hamburg; 1803; Heft 6: 312–321. 2 pl. (Pl.xxxi, xxxii).

NOTE. Translation of rec.112.

137. Description ostéologique du rhinocéros unicorne. *Paris. Muséum national d'histoire naturelle. Annales;* 1804; 3: 32–52. 4 pl.

NOTE. Also published in *Recherches sur les ossemens fossiles* . . . 1812, (rec.740), v.2 Mém.no.III. Translated into English, rec.166, and extracted in German, rec.168.

138. Description ostéologique du Tapir. *Paris. Muséum national d'histoire naturelle. Annales;* 1804; 3: 122–131.

NOTE. Published also in *Recherches sur les ossemens fossiles* . . . 1812, (rec.740), v.2 Mém.no.VII.

139. Sur quelques dents et os trouvés en France, qui paroissent avoir appartenu à des animaux du genre Tapir. *Paris. Muséum national d'histoire naturelle. Annales;* 1804; 3: 132–143. 5 pl.

NOTE. Also published in *Recherches sur les ossemens fossiles* . . . 1812, (rec.740), v.3 Mém no.VIII. Additions and corrections appeared in rec.151.

140. Description ostéologique et comparative du Daman. Hyrax Capensis. *Paris. Muséum national d'histoire naturelle. Annales;* 1804; 3: 171–182. 2 pl.

NOTE. Published also in *Recherches sur les ossemens fossiles* . . . 1812, (rec.740), v.2 Mém.no.II.

141. Sur les espèces d'animaux dont proviennent les os fossiles. Répandus dans la pierre à plâtre des environs de Paris. Premier Mémoire. Restitution de la tête. *Paris. Muséum national d'histoire naturelle. Annales;* 1804; 3: 275–303. 7 pl.

NOTE. Read at least in part at the September 5, 1803 meeting of the Académie des sciences. Published also in *Recherches sur les ossemens fossiles* . . . 1812, (rec.740), v.3 Mém.no.I. Continued by rec.142.

142. Suite des recherches Sur les os fossiles de la pierre à plâtre des environs de Paris par G. Cuvier. Deuxième Mémoire. Examen des dens et des portions de têtes éparses dans nos carrières à plâtre, qui diffèrent du *Palæotherium medium,* soit par l'espèce, soit même par le genre. *Paris. Muséum national d'histoire naturelle. Annales;* 1804; 3: 364–387. 6 pl.

NOTE. Presented at the October 3, 1803 session of the Académie des sciences. Published also in *Recherches sur les ossemens fossiles* . . . 1812, (rec.740), v.3 Mém.no.II, p.29–52. Preceded by rec.141 and succeeded by rec.143.

143. Suite des recherches Sur les os fossiles de la pierre à plâtre des environs de Paris. Par G. Cuvier. Troisième mémoire. Restitution des pieds. *Paris. Muséum national d'histoire naturelle. Annales;* 1804; 3: 442–472. 6 pl.

NOTE. Possibly the paper read by Cuvier at the December 3, 1803 meeting of the Académie des sciences. Preceded by rec.142 and succeeded by rec.181. Published also in *Recherches sur les ossemens fossiles* . . . 1812, (rec.740), v.3 Mém.no.III, p.53–82.

144. Suite des recherches sur les os fossiles de la pierre à plâtre des environs de Paris. Cinquième mémoire. (1) Sur les os du tronc. Première section. Description d'un squelette presque entier trouvé dans les carrières de Pantin. *Paris. Muséum national d'histoire naturelle. Annales;* 1804; 4: 66–75. 1 pl.

NOTE. Read at the Académie des sciences, Dec. 4, 1803. Continuation of rec.221 and continued by rec.222. Published also in *Recherches sur les ossemens fossiles* . . . 1812, (rec.740), v.3 Mém.no.V, p.1–10. A brief report was published, rec.161, which was translated into German, rec.169 and rec.189.

145. Mémoire sur l'Ibis des Anciens Egyptiens. *Paris. Muséum national d'histoire naturelle. Annales;* 1804; 4: 116–135. 3 pl.

NOTE. Published separately, n.p., n.d., 20 p., 3pl. Slightly expanded version of rec.68 which was translated into English in 1800, rec.78. Summarized, rec.72. Also published in a number of different editions of *Discours sur les révolutions . . . du globe* and *Recherches sur les ossemens fossiles,* which were not compared with the above.

146. Mémoire concernant l'animal de l'Hyale, un nouveau genre de mollusques nus, intermédiaire entre l'Hyale et le Clio, et l'établissement d'un nouvel ordre dans la classe des mollusques. *Paris. Muséum national d'histoire naturelle. Annales;* 1804; 4: 223–234. 1 pl.

NOTE. Also published with minor differences in *Mémoires pour servir à l'histoire et à l'anatomie des mollusques,* (rec.724), no.3.

147. Sur l'Hippopotame et sur son ostéologie. *Paris. Muséum national d'histoire naturelle. Annales;* 1804; 4: 299–328. 3 pl.

NOTE. Additions and corrections to this article appear in rec.152. Also published in *Recherches sur les ossemens fossiles* . . . 1812 (rec.740), v.2 Mém.no.V. The latter includes the additions and corrections of rec.152. Translated into German with substantial omissions, rec.171.

148. Mémoire Sur les Thalides (*Thalia,* Brown.), et sur les *Biphores* (*Salpa* Forskaohl). *Paris. Muséum national d'histoire naturelle. Annales;* 1804; 4: 360–382. 1 pl.

NOTE. Also published in *Mémoires pour servir à l'histoire et à l'anatomie des mollusques,* (rec.724), no.19. A German translation was published in 1820, rec.371.

149. Mémoire sur le genre Doris. *Paris. Muséum national d'histoire naturelle. Annales;* 1804; 4: 447–473. 2 pl.

NOTE. Also published in *Mémoires pour servir à l'histoire et à l'anatomie des mollusques,* (rec.724), no.5.

150. Mémoire sur l'Onchidie, genre de mollusques nuds, voisin des Limaces, et sur une espèce nouvelle, *Onchidium peronii. Paris. Muséum national d'histoire naturelle. Annales;* 1804; 5: 37–51. 1 pl.

NOTE. Also published in *Mémoires pour servir à l'histoire et à l'anatomie des mollusques,* (rec.724), no.13, with slight differences.

151. Additions et corrections à l'article sur les ossemens fossiles du Tapir. *Paris. Muséum national d'histoire naturelle. Annales;* 1804; 5: 52–55.

NOTE. Additions to rec.139. Also published and somewhat expanded in *Recherches sur les ossemens fossiles*. . . 1812, (rec.740), v.2 Mém. no.VIII.

152. Additions à l'article sur l'hippopotame. *Paris. Muséum national d'histoire naturelle. Annales;* 1804; 5: 56–57.

NOTE. Additions to rec.147.

153. Sur les ossemens fossiles d'Hippopotame. *Paris. Muséum national d'histoire naturelle. Annales;* 1804; 5: 99–122. 3 pl.

NOTE. Published also in *Recherches sur les ossemens fossiles* . . . 1812, (rec.740), v.2 Mém.no.VI.

154. Observations sur l'ostéologie des Paresseux. *Paris. Muséum national d'histoire naturelle. Annales;* 1804; 5: 189–215. 4 pl.

NOTE. Also published in *Recherches sur les ossemens fossiles* . . . 1812, (rec.740), v.4 Ptie.4 Mém.no.VI.

155. Mémoire Sur la Phyllidie et sur le Pleuro-branche, deux nouveaux genres de mollusques de l'ordre des gastéropodes, et voisins des patelles et des oscabrions, dont l'un est nu et dont l'autre porte une coquille cachée. *Paris. Muséum national d'histoire naturelle. Annales;* 1804; 5: 266–276. 1 pl.

NOTE. Also published in *Mémoires pour servir à l'histoire et à l'anatomie des mollusques,* (rec.724), no.8. Continued by rec.185.

156. Mémoire Sur le squelette presque entier d'un petit quadrupède du genre des Sarigues, trouvé dans la pierre à plâtre des environs de Paris. *Paris. Muséum national d'histoire naturelle. Annales;* 1804; 5: 277–292. 1 pl.

NOTE. Possibly the paper read by Cuvier at the December 31, 1804 meeting of the Académie des sciences. Also published in *Re-*

cherches sur les ossemens fossiles . . . 1812, (rec.740), v.3 Ptie.2 Mém.no.X. An extract was published in 1805, rec.179, and a condensation in German, rec.205.

157. Sur le Mégalonix, Animal de la famille des Paresseux, mais de la taille du Boeuf, dont les ossemens ont été découverts en Virginie, en 1796. *Paris. Muséum national d'histoire naturelle. Annales;* 1804; 5: 358–375. 1 pl.

NOTE. Published also in *Recherches sur les ossemens fossiles* . . . 1812, (rec.740), v.4 Ptie.4 Mém.no.VII. Summarized in German, rec.192.

158. Sur le Megatherium Autre animal de la famille des *Paresseux,* mais de la taille du *Rhinocéros,* dont un squelette fossile presque complet est conservé au cabinet royal d'histoire naturelle à Madrid. *Paris. Muséum national d'histoire naturelle. Annales;* 1804; 5: 376–387. 2 pl.

NOTE. Appended to this article, on p. 387–400, is a report on the same specimen by D. Jean-Baptiste Bru, translated by Bonpland and abridged, "Description des os du MEGATHERIUM, faite en montant le squelette." Above was published also in *Recherches sur les ossemens fossiles* . . . 1812, (rec.740), v.4 Ptie.4 Mém.no.VIII, p.19–43. Cuvier text in latter appears on p.19–30 and Bru on p.30–43. Translated into German, rec.191. An extract of a report given by Cuvier to the Académie des sciences on the same specimen was published in 1796, rec.22. For more information on the publications of Cuvier on this specimen, see López Piñero, José M., "Juan Bautista Bru (1740–1799) and the Description of the Genus *Megatherium,*" *Journal of the history of biology;* 1988; 21: 147–163.

159. Sur la *Dolabelle,* sur la *Testacelle,* et sur un nouveau genre de mollusques à coquille cachée, nommée *Parmacelle. Paris. Muséum national d'histoire naturelle. Annales;* 1804; 5: 435–444. 1 pl.

NOTE. Also published in *Mémoires pour servir à l'histoire et à l'anatomie des mollusques,* (rec.724), no.12, with minor differences.

160. Recherches d'anatomie comparée sur les dents, par le C. Cuvier. *Société philomatique de Paris. Bulletin;* 1811 [i.e.1801–1805]; [Ser.I] 3(82): 165–168.

NOTE. Issue of Dec.,1803/Jan.,1804. Translated into German, rec.188; and also condensed in German, rec.170.

161. Notice sur le squelette fossile trouvé à Pantin, dans une carrière à plâtre. *Société philomatique de Paris. Bulletin;* 1811 [i.e.1801–1805]; [Ser.I] 3(85): 189–190. 1 pl.

NOTE. Issue of March/April, 1804. Translated twice into German, rec.169 and rec.189. Expanded into an article, rec.144.

162. Note sur l'estomac et le canal intestinal du Kanguroo-géant, et du Kanguroo-rat. *Société philomatique de Paris. Bulletin;* 1811 [i.e. 1801–1805]; [Ser.I] 3(89): 221–222.

NOTE. Issue of July/Aug., 1804.

163. Sur les animaux auxquels appartenoient les pierres dites *Nummulaires* ou *Lenticulaires,* et sur ceux des *cornes d'Ammon. Société philomatique de Paris. Bulletin;* 1811 [i.e.1801–1805]; [Ser.I] 3(91): 237–238.

NOTE. Issue of Sept./Oct.,1804. Translated into German with omissions and some slight differences (or translated from another reporting of the same presentation), rec.190.

164. Mémoires sur plusieurs genres de Mollusques. *Société philomatique de Paris. Bulletin;* 1811 [i.e.1801–1805]; [Ser.I] 3(92): 245–247. 1 pl.

NOTE. Issue of Oct./Nov., 1804. See also rec.165, rec.184, rec.185.

165. Suite de l'extrait des mémoires sur les Mollusques. *Société philomatique de Paris. Bulletin;* 1811 [i.e.1801–1805]; [Ser.I] 3(93): 254–256. 1 pl.

NOTE. Issue of Nov./Dec., 1804 See also rec.164, rec.184, and rec.185.

166. Osteological Description of the one-horned Rhinoceros, by Cuvier. *Philosophical magazine,* London; 1804; 19: 350–354. 1 pl.; Oct.1804/Jan.1805 (20): 111–120. 2 pl.

NOTE. Translation of rec.137.

167. Anatomische Bemerkungen über einige Species der Mollusken aus dem Aplysia=Geschlechte. (Vom Hrn. Cuvier . . .). *Magazin für den neuesten Zustand der Naturkunde,* Jena; 1804; 7: 193–197.

NOTE. Translation of rec.126, but not compared. Rec.186 is a different German translation of the same article.

168. Osteologische Beschreibung des einhörnigen Rhinozeros von Cuvier. . . . *Magazin für den neuesten Zustand der Naturkunde,* Jena; 1804; 7: 215–232. 1 pl.

NOTE. German extract of rec.137.

169. Beschreibung eines fossilen Gerippes vom *Palaeotherium* welches in den Gipsbrüchen von Pantin gefunden worden. Vom Hrn. Cuvier. *Magazin für den neuesten Zustand der Naturkunde,* Jena; 1804; 8: 193–195. 1 pl.

NOTE. Translation of rec.161. Rec.189 is another German translation.

170. Vergleichbar=anatomische Untersuchungen über die Zähne; vom Hrn. Cuvier (A. d. Schr. des Soc. Philom.). *Magazin für den neuesten Zustand der Naturkunde,* Jena; 1804; 8: 272–284.

NOTE. A condensed report in German of rec.160. Translated in full in German, rec.188.

171. Über das Nilpferde und seinen Knochenbau. *Magazin für den neuesten Zustand der Naturkunde,* Jena; 1804; 8: 449–461. 1 pl.

NOTE. Translation with substantial omissions of rec.147.

172. Abhandlung über das Geschlecht Laplysia, gewöhnlich Giftkuttel (*Lepus marinus,* lièvre marin) genannt, nebst einigen Speciebus und der Anatomie desselben von G. Cuvier. Mit vier Kupfertafeln. *Paris. Mu-*

séum national d'histoire naturelle. Annalen des Nationalmuseums der Naturgeschichte, Hamburg; 1804; Heft 10: 249–274. 4 pl. (Pl.52–55).

NOTE. Translation of rec.128.

173. Critische Abhandlung über diejenige Species von Krebsen, welche den Alten schon bekannt war; so wie über die Namen, welche sie ihnen gaben, von G. Cuvier. *Paris. Muséum national d'histoire naturelle. Annalen des Nationalmuseums der Naturgeschichte,* Hamburg; 1804; Heft 11: 320–334.

NOTE. Translation of rec.129.

174. Nachricht über das Anlegen der Sammlung für vergleichende Anatomie im naturhistorischen Museum von G. Cuvier. *Paris. Muséum national d'histoire naturelle. Annalen des Nationalmuseums der Naturgeschichte,* Hamburg; 1804; Heft 11: 357–362.

NOTE. Translation of rec.127.

175. Rapport . . . sur une notice manuscrite de l'histoire naturelle du Petit Nord de Terre-Neuve . . . [par] M. Rivaud de St Germain. *Académie des sciences. Procès verbaux;* 1913; 3(1804–1807): 172–173.

NOTE. Read by Cuvier in his name and that of Lacépède at the Jan. 7, 1805 meeting of the Académie. Neither was indicated as reporter.

176. Rapport . . . sur le Mémoire de M.M. Péron et Lesueur, relatif au tablier des femmes hottentotes. *Académie des sciences. Procès verbaux;* 1913; 3(1804–1807): 197–198.

NOTE. Presented at the meeting of March 25, 1805. Signed by La Billardière and Cuvier as reporter.

177. Rapport . . . sur le Mémoire de M. Duvernoy, sur l'hymen. *Académie des sciences. Procès verbaux;* 1913; 3(1804–1807): 244–245.

NOTE. Signed by Portal and Cuvier as reporter. Read before the Académie, August 8, 1805.

178. Rapport . . . sur l'ouvrage de MM. Daudebart de Férussac, intitulé *Exposé d'un système conchyliogique. Académie des sciences. Procès verbaux;* 1913; 3(1804–1807): 285–286.

NOTE. Presented at the December 16, 1805 session of the Académie by Lamarck and Cuvier as reporter. Also published in: Férussac, Jean Baptiste, *Essai d'une methode conchyliologique, Appliquée aux Mollusques* . . . . Par M. Daudebard de Férussac, . . . Nouvelle éditon. . . . terminé par J. Daudebard fils. Paris, Delance, 1807, p. [v]–x. The latter indicates that Cuvier is the reporter. In some sources the *Essai*. . . is attributed to André Férussac.

179. Mémoire sur le squelette presque entier d'un petit quadrupède du genre des *Sarigues,* trouvé dans la pierre à plâtre des environs de Paris; . . . Extrait. *Journal de physique;* 1805; 61: 39–45. 1 pl.

NOTE. Condensation of rec.156. The above extract was further extracted and condensed in German, rec.205.

180. Sur les ossemens fossiles d'Hyènes. *Paris. Muséum national d'histoire naturelle. Annales;* 1805; 6: 127–144. 1 pl.

NOTE. Read before the Académie des sciences on March 25, 1805. Published also in *Recherches sur les ossemens fossiles* . . . 1812, (rec.740), v.4 Ptie 4 Mém.no.II. A summary of this paper appeared in 1808, rec.251.

181. Troisième mémoire, . . . [sur les os fossiles des environs de Paris]. IIe. Section. Restitution des pieds de devant. *Paris. Muséum national d'histoire naturelle. Annales;* 1805; 6: 253–283. 5 pl.

NOTE. Continuation of rec.143. Continued by rec.218. Published also in Recherches sur les ossemens fossiles . . . 1812, (rec.740), v.3 Mém.no.III, p.83–114.

182. Sur des ossemens fossiles trouvés en divers endroits de France, et plus ou moins sembables à ceux du Palæotherium. *Paris. Muséum national d'histoire naturelle. Annales;* 1805; 6: 346–366. 2 pl.

NOTE. Also published in *Recherches sur les ossemens fossiles. . .* 1812, (rec.740), v.3 Pt.2 Mém. no.VIII.

183. Mémoire Sur la *Scyllée, l'Éolide,* et le *Glaucus,* avec des additions au Mémoire sur la Tritonie. *Paris. Muséum national d'histoire naturelle. Annales;* 1805; 6: 416–436. 1 pl.

NOTE. Read at the August 19, 1805 meeting of the Académie des sciences. Also published in *Mémoires pour servir à l'histoire et à l'anatomie des mollusques,* (rec.724), no.6.

184. Suite de l'extrait des mémoires sur les Mollusques, . . . contenant la partie anatomique. *Société philomatique de Paris. Bulletin;* 1811 [i.e.1801–1805]; [Ser.I] 3(94): 261–263.

NOTE. Issue of Dec./Jan., 1804/1805. See also rec.164, rec.165, and rec.185.

185. Suite des mémoires sur les Mollusques, . . . sur les genres Phyllidie et Pleuro-branche. *Société philomatique de Paris. Bulletin;* 1811 [i.e.1801–1805]; [Ser.I] 3(96): 277–278.

NOTE. Issue of Feb./March, 1805. Continuation of rec.155. See also rec.164, rec.165, rec.184.

186. Nachricht von der Zergliederung einiger Aplysienarten von Cuvier. *Archiv für Zoologie und Zootomie,* Braunschweig; 1805; 4(2): 180–183.

NOTE. Translation of rec.126. A different German translation was also published, rec.167.

187. Ueber die *Pennatula cynomorium* (*Alcyonium epipetrum Gmel.*) und über die Korallen überhaupt, gelesen in der philomatischen Gesellschaft von Cuvier. *Archiv für Zoologie und Zootomie,* Braunschweig; 1805; 4(2): 190–191.

NOTE. Translation of rec.130.

188. Vergleichende anatomische Untersuchungen über die Zähne, von Cuvier gelesen in der philom. Gesellschaft. *Archiv für Zoologie und Zootomie,* Braunschweig; 1805; 4(2): 205–214.

NOTE. Translation of rec.160. A condensation in German also appeared, rec.170.

189. Ueber das zu Pantin in einem Gypsbruche gefundene fossile Knochengerüst, von Cuvier. *Archiv für Zoologie und Zootomie,* Braunschweig; 1805; 4(2): 224–225.

NOTE. Translation of rec.161 with omission of plate and explanation. Another German translation was also published, rec.169.

190. Ueber die Thiere, welchen die sogenannten Pfennig= oder Linseensteine zugehören; so wie über die, wovon die Ammonshörner herrühren. Vom Hrn. Cuvier. *Magazin für den neuesten Zustand der Naturkunde,* Jena; 1805; 10: 181–185.

NOTE. German translation of rec.163 with omissions and slight differences, or a translation of another reporting of the same presentation.

191. Ueber das Megatherium; vom Herrn Cuvier. *Magazin für den neuesten Zustand der Naturkunde,* Jena; 1805; 10: 419–436. 1 pl.

NOTE. Translation of rec.158.

192. Ueber den Megalonix. Ein Quadruped aus der Familie der Faulthiere, aber von der Taille des Ochsen, dessen Knochen in Virginien 1796 entdeckt worden find. (Vom Hrn. Cuvier, . . .). *Magazin für den neuesten Zustand der Naturkunde,* Jena; 1805; 10: 481–490. 1 pl.

NOTE. German summary of rec.157.

193. Rapport . . . sur la pièce anatomique faite pour l'École de Médecine par M. Laumonier. *Académie des sciences. Procès verbaux;* 1913; 3(1804–1807): 345–346.

NOTE. Presented at the meeting of the Académie on April 28, 1806. Signed by Sabatier, Fourcroy, Tenon, Hallé, Chaptal, and Cuvier as reporter.

194. Rapport . . . sur le voyage de M. Péron. *Académie des sciences. Procès verbaux;* 1913; 3(1804–1807): 363–367.

NOTE. Report read at the meeting of June 9, 1806 of the Académie des sciences and signed by Laplace, Bougainville, Fleurieu, Lacépède, and Cuvier as reporter. Published with omissions, rec.229. Condensed translation of latter was also published, rec.273.

195. *Traité élémentaire d'Histoire Naturelle* par A. M. Constant Duméril, . . . ; ouvrage composé par ordre du Gouvernement, pour servir à l'enseignement dans les Lycées. *Le Moniteur universel,* Paris; 1806 Nov. 22: 1408.

NOTE. Review by Cuvier.

196. Sur les Rhinocéros fossiles. *Paris. Muséum national d'histoire naturelle. Annales;* 1806; 7: 19–52. 4 pl.

NOTE. Also published in *Recherches sur les ossemens fossiles . . .* (rec.740), v.2 Mém.no.IV. Summarized with extracts in German, rec.206.

197. Mémoire Sur la Limace (*limax,* L.) et le Colimaçon (*helix,* L.). *Paris. Muséum national d'histoire naturelle. Annales;* 1806; 7: 140–184. 2 pl.

NOTE. Also published in *Mémoires pour servir à l'histoire et à l'anatomie des mollusques,* (rec.724), no. 11.

198. Mémoire Sur le Limnée (*Helix stagnalis,* Lin.) et le Planorbe (*Helix cornea,* ejusd.). *Paris. Muséum national d'histoire naturelle. Annales;* 1806; 7: 185–198. 1 pl.

NOTE. Also published in *Mémoires pour servir à l'histoire et à l'anatomie des mollusques,* (rec.724), no.14.

199. Sur les ossemens du genre de l'ours, qui se trouvent en grande quantité dans certaines cavernes d'Allemagne et de Hongrie. *Paris. Muséum national d'histoire naturelle. Annales;* 1806; 7: 301–372. 7 pl.

NOTE. Includes on p. 314–318 an article, "Examen d'analyse de la terre servant d'enveloppe aux os de la caverne de Gaylenreuth," by Laugier. Above was published also in *Recherches sur les ossemens fossiles . . .* 1812, (rec.740), v.4 Ptie. 4 Mém.no.I. Extracted and translated into German, rec.231.

200. Sur les éléphans vivans et fossiles. *Paris. Muséum national d'histoire naturelle. Annales;* 1806; 8: 1–58; 93–155 5pl.; 249–269.

NOTE. The above were presented as 3 memoirs read before the Académie des sciences on May 5, 12, and 26, 1806. They were also published in *Recherches sur les ossemens fossiles . . .* , 1812, (rec.740), v.2 Mém.no.IX. Two of the memoirs, p.1–58 and 93–155, were translated into English with some omissions: rec.204 and rec.230 respectively.

201. Sur le grand Mastodonte, Animal très-voisin de l'éléphant, mais à mâchelières hérissés de gros tubercules, dont on trouve les os en divers endroits des deux continens, et surtout près des bords de l'Ohio, dans l'Amérique Septentrionale, improprement nommé *Mammouth* par les Anglais et par les habitans des États-Unis. *Paris. Muséum national d'histoire naturelle. Annales;* 1806; 8: 270–312. 8 pl.

NOTE. Published also in *Recherches sur les ossemens fossiles . . .* 1812, (rec.740), v.2 Mém.no.X.

202. Sur différentes dents du genre des Mastodontes, mais d'espèces moindres que celles de l'Ohio, trouvèes en plusieurs lieux des deux continens. *Paris. Muséum national d'histoire naturelle. Annales;* 1806; 8: 401–420. 4 pl.

NOTE. Also published in *Recherches sur les ossemens fossiles . . .* 1812, (rec.740), v.2 Mém.no.XI.

203. Résumé général De l'Histoire des ossemens fossiles de *Pachydermes* des terrains meublés, et d'alluvion. *Paris. Muséum national d'histoire naturelle. Annales;* 1806; 8: 420–424.

NOTE. Published also with an additional final paragraph in *Recherches sur les ossemens fossiles . . .* 1812, (rec.740), v.2 Mém.no.XII.

204. Memoir upon living and fossil Elephants. By G. Cuvier. *Philosophical magazine,* London; Oct.1806/Jan.1807; 26: 158–169, 204–211, 302–313.

NOTE. Translation of p.1–58 of rec.200 with some omissions. Continued by rec.230.

205. Merkwürdige Resultate Cuvier's aus Untersuchungen fossiler Knochen. 1. Auszug eines Briefes von Herrn Cuvier in Paris an den geh. Oberbergrath Karsten in Berlin. 2. Aus einer Abhandlung Cuvier's über ein beinahe vollständiges Skelett eines kleinen vierfüssigen Thiers vom Geschlechte *Sarigue* aus den Gypsbrüchen bei Paris. *Annalen der Physik,* Halle; 1806; 23: 109–113.

NOTE. Cuvier's letter of Pt.1 is dated 9 May, 1806. Pt.2 is a translated extract and condensation of rec.179.

206. Ueber die fossilen Rhinocer; vom Hrn. Cuvier. *Magazin für den neuesten Zustand der Naturkunde,* Jena; 1806; 12: 97–113. 1 pl.

NOTE. Summary with extracts of rec.196.

207. Rapport à MM. les professeurs administrateurs du Muséum d'histoire naturelle, Sur les collections d'histoire naturelle que M. Leschenault de la Tour a rapportées de son voyage à Java, de 1803 à 1807. [Paris: Imprimerie de A. Belin; 180?]. 8p. 8°.

NOTE. Report presented by Cuvier and signed by Lamarck, Desfontaines, and Cuvier at the Muséum d'histoire naturelle on Oct. 14, 1807, as requested by the Minister of the Interior.

208. Rapport sur un mémoire de M. de Candolle, intitulé *Tableau de la nutrition des végétaux,* Par MM. Chaptal, Labillardière et Cuvier . . . *Académie des sciences. Mémoires;* 1807; 8(1): 68–89.

NOTE. Presented at the December 9, 1805 meeting of the Académie and signed by Chaptal, La Billardière, and Cuvier as reporter. Also published in the *Académie des sciences, Procès verbaux;* 1913; 3(1804–1807): 277–284.

209. Rapport Sur un ouvrage manuscrit de M. André, ci-devant connu sous le nom de P. Chrysologue de Gy, lequel ouvrage est intitulé *Théorie de la surface actuelle de la terre,* Par MM. Hauy, Lelièvre, et Cuvier, rapporteur. Lu le 11 Août 1806. *Académie des sciences. Mémoires;* 1807; 8(1): 128–145.

NOTE. Also published in the *Journal des mines,* 1807; 21: 413–430; in the *Académie des sciences, Procès verbaux;* 1913; 3(1804–1807): 408–413; and translated into English, rec.272.

210. Rapport . . . sur un Mémoire de M. Geoffroy concernant la Nageoire pectorale des poissons. *Académie des sciences. Procès verbaux;* 1913; 3(1804–1807): 523–525.

NOTE. Presented at the meeting of May 4, 1807 and signed by Bosc, Lacépède, and Cuvier as reporter.

211. Rapport . . . sur les figures de pigeons présentées par Mlle. de Courcelles. *Académie des sciences. Procès verbaux;* 1913; 3(1804–1807): 539–540.

NOTE. Presented at the meeting of June 15, 1807 and signed by Lacépède, and Cuvier as reporter.

212. Rapport . . . sur le Mémoire de M. Descourtils relatif au *Crocodile de St. Dominge. Académie des sciences. Procès verbaux;* 1913; 3(1804–1807): 542–543.

NOTE. Presented at the meeting of June 15, 1807 and signed by Tenon, Lacépède, and Cuvier as reporter.

213. Rapport sur un Mémoire de M. Geoffroy relatif au Sternum des poissons et à ses annexes. *Académie des sciences. Procès verbaux;* 1913; 3(1804–1807): 564–565.

NOTE. Presented at the Aug. 24, 1807 meeting and signed by Tenon, Lacépède, and Cuvier as reporter.

214. Rapport sur le Mémoire de M. Geoffroy . . . intitulé: Considérations sur les pièces de la tête osseuse des animaux vertébrés et particulièrement sur celles du crâne des oiseaux. *Académie des sciences. Procès verbaux;* 1913; 3(1804–1807): 575–576.

NOTE. Presented at the meeting of Sept. 7, 1807 and signed by Lamarck, Lacépède, and Cuvier as reporter.

215. Sur les espèces des animaux carnassiers dont on trouve les ossemens mêlés à ceux d'ours, dans les cavernes d'Allemagne et de Hongrie; . . . Extrait. *Journal de physique;* 1807; 65: 282–289.

NOTE. Shortened version of rec.224 with minor changes. Translated into English, rec.271.

216. Réponse de M. Cuvier à M. Rampasse. *Journal de physique;* 1807; 65: 430–431.

NOTE. Cuvier identifies bones sent from Corsica as *Lagomys.* Translated into English, rec.270.

217. Traité élémentaire de minéralogie; avec des applications aux arts; Ouvrage destiné à l'enseignement dans les Lycées nationaux; par Alexandre Brongniart, . . . 2v. in 8°. Paris, Déterville – 1807. Extrait. *Le Moniteur universel,* Paris; 1807 November 3; (307): 1186–1188.

NOTE. Article signed by Cuvier. Also published separately, n.p., n.d., 19p. Published again in 1885, rec.562, with the addition of a letter from Cuvier to M. [Claude] Lecoz, Archevèque de Besançon, analyzing Brongniart's *Traité. . . .*

218. Suite des recherches sur les os fossiles des environs de Paris. . . . Troisième Mémoire. Troisième Section. Les phalanges. *Paris. Muséum national d'histoire naturelle. Annales;* 1807; 9: 10–15. 1 pl.

NOTE. Continuation of rec.181. Continued by rec.219. Published also in *Recherches sur les ossemens fossiles* . . . 1812, (rec.740), v.3 Mém.no.III, p.115–120.

219. [Suite des recherches sur les os fossiles des environs de Paris.] Quatrième Mémoire. Sur les os longs des extrémités. Première Section. Les os des extrémités postérieures. *Paris. Muséum national d'histoire naturelle. Annales;* 1807; 9: 16–44. 5 pl.

NOTE. Continuation of rec.218 and continued by rec.220. Published also in *Recherches sur les ossemens fossiles* . . . 1812, (rec.740), v.3 Mém.no.IV, p.121–150.

220. Suite des recherches sur les os fossiles des environs de Paris.... Quatrième Mémoire. Seconde Section. Les os longs des extrémités antérieures. *Paris. Muséum national d'histoire naturelle. Annales;* 1807; 9: 89–102. 2 pl.

NOTE. Continuation of rec.219 and continued by rec.221. Published also in *Recherches sur les ossemens fossiles* . . . 1812, (rec.740), v.3 Mém.no.IV, p.151–164.

221. Suite des recherches sur les os fossiles des environs de Paris. . . . Quatrième Mémoire. Troisième Section. Les omoplates et les bassins. *Paris. Muséum national d'histoire naturelle. Annales;* 1807; 9: 205–215. 2 pl.

NOTE. Continuation of rec.220 and continued by rec.144. Published also in Recherches sur les ossemens fossiles . . . 1812, (rec.740), v.3 Mém.no.IV, p.165–174.

222. Suite des recherches sur les os fossiles des environs de Paris.... Ve. Mémoire. IIe. Section. Description de deux squelettes presque entiers d'*Anoplotherium commune. Paris. Muséum national d'histoire naturelle. Annales;* 1807; 9: 272–282. 2 pl.

NOTE. Possibly the memoir or memoirs read at the March 9 and or April 13, 1807 meetings of the Académie des sciences. Continuation of rec.144 and continued by rec.248. Also published in *Recherches sur les ossemens fossiles* . . . 1812, (rec.740), v.3 Mém.no.V, p.11–22.

223. Mémoire Sur les ossemens d'oiseaux qui se trouvent dans les carrières de pierres à plâtre des environs de Paris. *Paris. Muséum national d'histoire naturelle. Annales;* 1807; 9: 336–356. 2 pl.

NOTE. Continued by rec.259. Published also in *Recherches sur les ossemens fossiles* . . . 1812, (rec.740), v.3 Mém.no.XI.

224. Sur les espèces des animaux carnassiers Dont on trouve les os mêlés à ceux d'ours dans les cavernes d'Allemagne et de Hongrie. *Paris. Muséum national d'histoire naturelle. Annales;* 1807; 9: 428–447. 2 pl.

NOTE. Published also in *Recherches sur les ossemens fossiles* . . . 1812, (rec.740), v.4 Ptie.IV Mém.no.III. A shortened version, rec.215, was also published and translated into English, rec.271.

225. Sur les différentes espèces de crocodiles vivans et sur leurs caractères distinctifs. *Paris. Muséum national d'histoire naturelle. Annales;* 1807; 10: 8–66. 2 pl.

NOTE. Also published in *Recherches sur les ossemens fossiles* . . . 1812, (rec.740), v.4 Ptie.5 Mém.no.1. Summarized in rec.228.

226. Mémoire sur quelques ossemens de carnassiers, épars dans les carrières à plâtre des environs de Paris. *Paris. Muséum national d'histoire naturelle. Annales;* 1807; 10: 210–217. 1 pl.

NOTE. Also published in the *Recherches sur les ossemens fossiles* . . . 1812, (rec.740), v.3 Mém.no.IX with 4 plates and an addendum. In a footnote to the latter, Cuvier says he has added 2 more plates his brother had had engraved for another work. Continued by rec.261?

227. Rapport à la Classe des sciences physiques et mathématiques de l'Institut par M. Cuvier. *Paris. Muséum national d'histoire naturelle. Annales;* 1807; 10: 381–386.

NOTE. Also published in the *Académie des sciences, Procès verbaux;* 1913; 3(1804–1807): 623–625, with the title, *Rapport . . . sur le MAMMOUTH découvert en Sibérie et décrit dans la Gazette de Berlin* [by the mineralogist, Karsten]. Presented at the Académie meeting of Nov. 30, 1807. Report also signed by Lacépède.

228. Mémoire sur les différentes espèces de Crocodiles vivans, et sur leurs caractères distinctifs. *Société philomatique de Paris. Bulletin;* 1807 [i.e.1807–1809]; [Ser. II] 1(1): 10–14.

NOTE. Appeared in issue of Oct., 1807. Summary of rec.225.

229. Rapport fait au gouvernement par l'Institut impérial, Sur le Voyage de Découvertes aux Terres Australes. Extrait du Procès-verbal de la Classe des Sciences physiques et mathématiques, séance du lundi 9 juin 1806. In: Péron, François. Voyage de découvertes aux terres australes, . . . , Sur les Corvettes le Géographe, le Naturaliste, et la Goelette le Casuarina, Pendant les Années 1800, 1801, 1802, 1803 et 1804 . . . Paris: De l'Imprimerie impériale; 1807; 1: i–xv.

NOTE. Report signed by Laplace, Bougainville, Fleurieu, Lacépède; and Cuvier as reporter. Extracts from rec.194. Also published in Paris: Arthus Bertrand, 1824, 2nd edition [same text as above], p.1–18. Translated into English with omissions and some condensation, rec.273. Other editions were not seen.

230. Additional Memoir upon living and fossil Elephants. By M. Cuvier. *Philosophical magazine,* London; June–Sept. 1807; 28: 258–264, 359–366; Oct.,1807–Jan.,1808; 29: 52–65, 244–254; Feb.–May, 1808; 30: 15–25. 8 pl. tab.

NOTE. Translation of p.93–155 of rec.200. Continuation of rec.204.

231. Ueber die Gebeine aus der Bärengattung, die sich in grosser Menge in gewissen Höhlen Deutschlands und Ungarns finden. von G. Cuvier. Uebersetzt von A. F. Gehlen. *Journal für die Chemie und Physik,* Berlin; 1807; 3: 37–48.

NOTE. Translation of extracts from rec.199 including the section by Laugier.

232. Rapport . . . sur un Mémoire de M. Duméril sur la *division des reptiles batraciens en deux familles naturelles. Académie des sciences. Procès verbaux;* 1913; 4(1808–1811): 23–25.

NOTE. Presented at the meeting of Feb. 29, 1808 of the Académie and signed by Lacépède, and Cuvier as reporter.

233. Des os fossiles trouvés dans l'Amérique-septentrionale. *Journal de physique;* 1808; 67: 331–333.

NOTE. Report presented at the Oct. 10, 1808 meeting of the Académie des sciences by Lacépède, and Cuvier as reporter, on fossil bones sent to the Académie by Thomas Jefferson. Published also in the *Académie des sciences, Procès verbaux;* 1913; 4(1808–1811): 113–114, and with different introductory paragraphs in rec.238.

234. Essai sur la géographie minéralogique des environs de Paris. Par MM. Cuvier et Alex. Brongniart. *Journal des mines;* 1808; 23: 421–458.

NOTE. Another publication of rec.241 with several small omissions. See Notes in rec.241 and rec.284 for further publishing history.

235. Discours de M. Cuvier, . . . [sur le progrès des sciences depuis 1789]. *Le Moniteur universel,* Paris; 1808 February 8; (39): 157–160.

NOTE. Also published in *Académie des sciences, Mémoires;* 1808; 9(Hist): 203–229; France. Conseil-d'État, *Compte rendu à Sa Majesté l'EMPEREUR ET ROI, par l'Institut, de l'état des Sciences, des Lettres et des Arts, et de leurs progrès depuis 1789, dans la séance du Conseil-d'État, du samedi 6 février 1808. Première classe,* [n.p.; n.d.]: 26–45; and in Holland, rec.732. A greatly expanded edition based on above was published in 1810, rec.733, and added volumes later extended the years covered. (See Note to rec.694.)

236. Extrait d'un rapport sur un Mémoire de MM. Gall et Spurzheim, relatif à l'anatomie du cerveau—Séance du 25 avril et 2 mai [1808]. *Le Moniteur universel,* Paris; 1808 June 20; (172): 679–680.

NOTE. Extracts from rec.242. Report signed by Tenon, Portal, Sabatier, Pinel, and Cuvier as reporter.

237. Tableaux de la nature, ou Considérations sur les déserts, sur la physionomie des végétaux et sur les cataractes de l'Oréonoque; par A. de Humboldt, traduits de l'allemand par J. B. B. Eyries. *Le Moniteur universel,* Paris; 1808 July 25; (207): 818–819.

NOTE. Review signed, G. Cuvier. The volume reviewed was published; Paris, Schoell, 1808. A note at end of review announces that a following section of the review would be published later, but a continuation was not found.

238. [Rapport sur une collection d'os fossiles de l'Amérique septentrionale envoyée par M. Jefferson]. *Le Moniteur universel,* Paris; 1808 October 14; (288): col.1135.

NOTE. Report made to the Académie des sciences on October 10 by Lacépède and Cuvier as reporter. Also published with different introductory paragraphs, rec.233.

239. Mémoire Sur la Janthine et sur la Phasianelle de M. Lamarck. *Paris. Muséum national d'histoire naturelle. Annales;* 1808; 11: 121–135. 1pl.

NOTE. Published also in *Mémoires pour servir à l'histoire et à l'anatomie des mollusques,* (rec.724), no.15.

240. Mémoire sur la Vivipare d'eau douce (*Cyclosloma* [sic] *viviparum* Draparn., *Helix vivipara* Lin.); sur quelques espèces voisines, et idée générale sur la tribu des gastéropodes pectinés à coquille entière. *Paris. Muséum national d'histoire naturelle. Annales;* 1808; 11: 170–188. 1 pl.

NOTE. Also published in *Mémoires pour servir à l'histoire et à l'anatomie des mollusques,* (rec.724), no.16. Summarized in rec.252.

241. Essai sur la géographie minéralogique des environs de Paris. Par MM. G. Cuvier et Alex. Brongniart. *Paris. Muséum national d'histoire naturelle. Annales;* 1808; 11: 293–326.

NOTE. Also published with several small omissions, rec.234. Translated into English, rec.281. Above is an expanded publication of what was probably an earlier version without date, rec.690. It was further expanded, rec.284 and rec.691.

242. Rapport fait à L'Institut Sur un Mémoire de MM. docteurs Gall et Spurzheim, [relatif à l'anatomie du cerveau]. *Paris. Muséum national d'histoire naturelle. Annales;* 1808; 11: 329–375.

NOTE. Read at April 18 and 25, 1808, sessions of the Académie des sciences, and presented by Tenon, Portal, Sabatier, Pinel and Cuvier as reporter. First published as extracts in *Le Moniteur universel,* rec.236. Published also with minor changes in wording in the *Académie des sciences, Mémoires;* 1808(Hist): 109–160, published in 1809; also in the *Journal de physique;* 1808; 67: 233–266; in the *Académie des Sciences, Classe des sciences mathématiques et physiques, Extrait des registres, séances des 25 avril et 2 mai 1808,*

Baudouin, n.p., n.d., 52p. 4°; and in *Académie des sciences, Procès verbaux;* 1913; 4(1808–1811): 48–64. Translated into English with numerous derogatory insertions, rec.269.

243. Mémoire Sur le grand Buccin de nos côtes (*Buccinum undatum,* LIN.) et son anatomie. *Paris. Muséum national d'histoire naturelle. Annales;* 1808; 11: 447–457. 1 pl.

NOTE. Published also in *Mémoires pour servir à l'histoire et à l'anatomie des mollusques,* (rec.724), no.17.

244. Observations sur l'ostéologie des crocodiles vivans. *Paris. Muséum national d'histoire naturelle. Annales;* 1808; 12: 1–26. 2 pl.

NOTE. Also published in *Recherches sur les ossemens fossiles . . .* 1812, (rec.740), v.4 Ptie.V Mém.no.II. A German condensation covering only the introduction, skull, and lower jaw appeared in 1818, rec.345.

245. Sur les ossemens fossiles de crocodiles, Et particulièrement sur ceux des environs du Havre et de Honfleur, avec des remarques sur les squelettes des sauriens de la Thuringe. *Paris. Muséum national d'histoire naturelle. Annales;* 1808; 12: 73–110. 2 pl.

NOTE. Probably the paper read on July 11, 1808 at the Académie des sciences. Also published in *Recherches sur les ossemens fossiles . . .* 1812, (rec.740), v.4 Ptie.V Mém.no.III.

246. Sur le grand animal fossile des carrières de Maestricht. *Paris. Muséum national d'histoire naturelle. Annales;* 1808; 12: 145–176. 2 pl.

NOTE. Read by Cuvier at the Aug. 1, 1808 meeting of the Académie. Published also in *Recherches sur les ossemens fossiles . . .* 1812, (rec.740), v.4 Ptie.V Mém.no.IV.

247. Mémoire sur le genre *Thethys* et sur son anatomie. *Paris. Muséum national d'histoire naturelle. Annales;* 1808; 12: 257–270. 1 pl.

NOTE. Also published in *Mémoires pour servir à l'histoire et à l'anatomie des mollusques,* (rec.724), no.7, with minor changes.

248. Suite des recherches sur les os fossiles des environs de Paris. Cinquième Mémoire, Troisième Section. Vertèbres et côtes isolées. *Paris. Muséum national d'histoire naturelle. Annales;* 1808; 12: 271–284. 2 pl.

NOTE. Continuation of rec.222. Published also in *Recherches sur les ossemens fossiles* . . . 1812, (rec.740), v.3 Mém.no.V.

249. Sur les os fossiles de ruminans, trouvés dans les terrains meubles. *Paris. Muséum national d'histoire naturelle. Annales;* 1808; 12: 333–398. 3 pl.

NOTE. Published also in Recherches sur les ossemens fossiles . . . 1812, (rec.740), v.4 Ptie.3 Mém.no.I. and extracted, rec.254. A partial summary, rec.264, was continued by rec.267, and was translated into German, rec.274.

250. *Sur le genre Glaucus. Société philomatique de Paris. Bulletin;* 1807 [i.e.1807–1809]; [Ser. II] 1(8): 148.

NOTE. Appeared in issue of May, 1808.

251. Note sur les ossemens fossiles d'hyènes. *Société philomatique de Paris. Bulletin;* 1807 [i.e.1807–1809]; [Ser. II] 1(9): 149–150.

NOTE. Appeared in issue of June, 1808. Summary of rec.180.

252. Sur la Vivipare d'eau douce *(Cyclostoma viviparum,* Drap.), et sur la tribu des Gastéropodes pectinés à coquille entière. *Société philomatique de Paris. Bulletin;* 1807 [i.e.1807–1809]; [Ser.II] 1(10): 166–167.

NOTE. Appeared in issue of July, 1808. Summary of rec.240.

253. Discours prononcé . . . devant les professeurs de l'Académie de Pise le xxi décembre MDCCCIX. Au nom de la commission chargée de l'inspection extraordinaire de l'instruction publique dans les départements italiens de l'Empire. Pise: Imprimerie de la Société Littéraire; 1809. [6]p. 25.5cm.

254. Sur les os fossiles de ruminans trouvés dans les terrains meubles; . . . Extrait. *Journal de physique;* 1809; 68: 358–377.

NOTE. An extract of rec.249.

255. Sur les brèches osseuses qui remplissent les fentes de rochers à Gibraltar et dans plusieurs autres lieux des côtes de la Méditerranée, et sur les animaux qui en ont fourni les os. *Paris. Muséum national d'histoire naturelle. Annales;* 1809; 13: 169–206. 2 pl.

NOTE. Read by Cuvier at the Académie des sciences meetings of March 27 and April 10, 1809. Also published in *Recherches sur les ossemens fossiles* . . . 1812, (rec.740), v.4 Ptie.3 Mém.no.II. A synopsis was published, rec.268, and a German translation, rec.394.

256. Sur l'ostéologie du Lamantin, sur la place que le *Lamantin* et le *Dugong* doivent occuper dans la méthode naturelle, et sur les os fossiles de *Lamantins* et *Phoques*. *Paris. Muséum national d'histoire naturelle. Annales;* 1809; 13: 273–312. 1 pl.

NOTE. Also published in *Recherches sur les ossemens fossiles* . . . 1812, (rec.740), v.4 Ptie. 4 Mém.no.IX and separately, n.p., n.d., 40 p., 1 pl. Synopsis was published, rec.265.

257. Sur quelques quadrupèdes ovipares fossiles conservés dans les schistes calcaires. *Paris. Muséum national d'histoire naturelle. Annales;* 1809; 13: 401–437. 2 pl.

NOTE. Probably the paper read by Cuvier at the June 19, 1809 meeting of the Académie des sciences. Also published in *Recherches sur les ossemens fossiles* . . . 1812, (rec.740), v.4 Ptie.5 Mém.no.V. Summary published, rec.266.

258. Des os fossiles de chevaux et de sangliers. *Paris. Muséum national d'histoire naturelle. Annales;* 1809; 14: 33–42.

NOTE. Published also in *Recherches sur les ossemens fossiles* . . . 1812, (rec.740), v.4 Ptie. 3 Mém.no.III. Translated into English, rec.282.

259. Supplément au mémoire sur les Ornitholithes de nos carrières à plâtre. *Paris. Muséum national d'histoire naturelle. Annales;* 1809; 14: 43–46. 1 pl.

NOTE. Continuation of rec.223. Also published in *Recherches sur les ossemens fossiles* . . . 1812, (rec.740), v.3 Mém.no.XI.

260. De quelques rongeurs fossiles, principalement du genre *Castors,* qui se sont trouvés dans des tourbes ou dans des alluvions, et de quelques autres rongeurs enfermés dans des schistes. *Paris. Muséum national d'histoire naturelle. Annales;* 1809; 14: 47–55. 1 pl.

NOTE. Published also in *Recherches sur les ossemens fossiles* . . . 1812, (rec.740), v.4 Ptie 4 Mém.no.V.

261. Recherches sur les espèces vivantes de grands chats, pour servir de preuves et d'éclaircissemens au chapitre sur les carnassiers fossiles. *Paris. Muséum national d'histoire naturelle. Annales;* 1809; 14: 136–164. 2 pl.

NOTE. Continuation of rec.226? Published also in *Recherches sur les ossemens fossiles* . . . 1812, (rec.740), v.4 Ptie.IV Mém.no.IV; and separately (n.p., n.d., 30p. 2pl.). Summarized in rec.280.

262. Rapport Fait à la classe des sciences physiques et mathématiques, sur le Mémoire de M. Delaroche, relatif à la vessie aérienne des poissons. Par G. Cuvier. *Paris. Muséum national d'histoire naturelle. Annales;* 1809; 14: 165–183.

NOTE. Published also in the *Académie des sciences, Procès verbaux;* 1913; 4(1808–1811): 246–251. Report signed by Lacépède, Vauquelin, and Cuvier. Translated into English, rec.283.

263. Sur les ossemens fossiles de tortues. *Paris. Muséum national d'histoire naturelle. Annales;* 1809; 14: 229–244. 2 pl.

NOTE. Also published in *Recherches sur les ossemens fossiles* . . . 1812, (rec.740), v.4 Ptie.6 Mém.no.VI, and as a separate (n.p., n.d., 16p.). An abstract was also published, rec.279.

264. Sur les Os fossiles des ruminans trouvés dans les terreins [sic] meubles. *Société philomatique de Paris. Bulletin;* 1807 [i.e.1807–1809]; [Ser.II] 1(21): 345–349.

NOTE. Issue of June, 1809. Published also in the *Journal des mines;* 1809; 26: 389–396. Summary of first part of rec.249 dealing with deer; balance of abstract dealing with cattle appears in rec.267. Translated into German, rec.274, with omission of one paragraph.

265. Sur les Lamantins et les Os fossiles de ces animaux. *Société philomatique de Paris. Bulletin;* 1807 [i.e. 1807–1809]; [Ser.II] 1(24): 395–396.

NOTE. Issue of Sept., 1809. Brief summary report of rec.256.

266. Sur quelques quadrupèdes ovipares fossiles conservés dans des schistes calcaires. *Société philomatique de Paris. Bulletin;* 1807 [i.e. 1807–1809]; [Ser.II] 1(24): 398–400.

NOTE. Issue of Sept., 1809. Summary of rec.257.

267. Suite de l'extrait du Mémoire de M. Cuvier, sur les Os fossiles de ruminans trouvés dans les terrains meubles. *Société philomatique de Paris. Bulletin;* 1807 [i.e.1807–1809]; [ser.II] 1(24): 400–402.

NOTE. Appeared in issue of Sept., 1809. Continuation of summary, rec.264, of Cuvier's memoir, rec.249. The continuation deals only with cattle.

268. Sur les brèches osseuses des côtes de la Méditerranée. *Société philomatique de Paris. Bulletin;* 1807 [i.e.1807–1809]; [Ser.II] 1(24): 402–404.

NOTE. Issue of Sept., 1809. Synopsis of rec.255.

269. Report on a Memoir of Drs. Gall and Spurzheim, relative to the anatomy of the brain. By M. M. Tenon, Portal, Sabatier, Pinel and Cuvier. With Critical Observations. By the Editor. *Archives of universal science,* Edinburgh; 1809 July; 3: 119–171.

NOTE. Translation of rec.242 with the addition of extensive and numerous derogatory insertions. Alexander Walker, "Lecturer in physiology," was editor and founder of the *Archives of universal science* and presumably was the author of the inserted comments. Cuvier text is in quotation marks.

270. Mr. Cuvier's answer to Mr. Rampasse. *Journal of natural philosophy,* London; 1809; 24: 196–197.

NOTE. Translation of rec.216.

271. Abridgment of a Paper on the Species of Carniverous Animals, the Bones of which are found mixed with those of Bears in Caverns in Germany and Hungary. By Mr. Cuvier. *Journal of natural philosophy,* London; 1809; 24: 295–302.

NOTE. Translation of rec.215.

272. Report on a manuscript work of M. André, formerly known under the name of P. Chrysologue de Gy, entitled *A theory of the actual surface of the earth.* By MM. Hauy, Levierre, and Cuvier. Read to the Class of Mathematical and Physical Sciences in the National Institute. *Philosophical magazine,* London; 1809; 33: 170–173, 312–320.

NOTE. Translation of rec.209.

273. Substance of the report to the French government by the Imperial Institute, on the voyage of discovery to the Southern hemisphere. In: Péron, François. A voyage of discovery to the Southern hemisphere, performed . . . during the years 1801, 1802, 1803, and 1804. . . . London: Richard Phillips; 1809: iii–viii. (Phillips, Sir Richard. A collection of modern and contemporary voyages and travels . . . ; v. XI (pt.1) 1810).

NOTE. Signed by Laplace, Bougainville, Fleurieu, Lacépède, and Cuvier as reporter. Translation of rec.229 with some condensation and omissions.

274. Ueber die fossilen Knochen wiederkäuender Thiere, die man in der Erde findet. *Bulletin des Neuesten und Wissenswürdigsten aus der Naturwissenschaft,* Berlin; 1809; 3: 218–226.

NOTE. Translation of rec.264 omitting one paragraph.

275. Discours . . . , aux Facultés de l'Académie de Gênes, Assemblées dans le Salon des receptions, le 26 Novembre 1809. Gênes: Imprimerie de l'Académie impériale; [1809?,1810?]. 8p. 26.5cm.

276. Discours [prononcé à l'Académie de Turin le IX avril MDCCCX par Monsieur Cuvier . . . en mission dans les départemens au-dela des alpes]. *Le Moniteur universel;* 1810 May 13; (133): 525–526.

NOTE. Also published separately, (n.p.; 1809?); 7p. Texts were not compared.

277. Mémoire sur les Acères, ou Gastéropodes sans tentacules apparens. *Paris. Muséum national d'histoire naturelle. Annales;* 1810; 16: 1–18. 1 pl.

NOTE. Also published separately, 18 p., 1 pl. and in *Mémoires pour servir à l'histoire et à l'anatomie des mollusques,* (rec.724), no.10, with 2 pl. The separate has an explanation for 2 plates, but the copy in the Muséum national d'histoire naturelle, Paris, had only 1 plate, the same plate as that in the above. An extract was also published, rec.286.

278. Mémoire sur les os de reptiles et de poissons des Carrières à plâtre des environs de Paris. *Paris. Muséum national d'histoire naturelle. Annales;* 1810; 16: 115–134. 1 pl.

NOTE. Read at the meeting of Oct. 15, 1810 of the Académie des sciences. Published also in *Recherches sur les ossemens fossiles . . .* 1812, (rec.740), v.3 Mém.no.XII. A preliminary summary was also published, rec.287.

279. Sur les tortues fossiles. *Société philomatique de Paris. Bulletin;* 1810, 1811; [Ser.II] 2(28): 7–8.

NOTE. Issue of Jan., 1810. Abstract of rec.263. Also published in *Journal des mines;* 1810; 27: 79–80. A slightly condensed German translation also was published, rec.295.

280. Recherches sur les espèces vivantes de grands Chats, pour servir de preuves et d'éclaircissement au chapitre sur les carnassiers fossiles.

Société philomatique de Paris. Bulletin; 1810, 1811; [Ser.II] 2(30): 45–49.

NOTE. Appeared in the issue of March, 1810. Summary of rec.261.

281. Memoir on the Mineralogical Geography of the environs of Paris. By Messrs. Cuvier and Brongniart. *Philosophical magazine,* London; Jan./June,1810; 35: 36–58.

NOTE. Translation of rec.241.

282. On the Fossil Bones of Horses and Wild Boars. By. G. A.[sic] Cuvier. *Philosophical magazine,* London; Jan./June, 1810; 35: 215–221.

NOTE. Translation of rec.258.

283. Report made to the French Institute on a Memoir of M. Delaroche on the Air-bladder of Fishes. By G. Cuvier. *Philosophical magazine,* London; Jan./June,1810; 35: 291–302.

NOTE. Translation of rec.262.

284. Essai sur la Géographie minéralogique des environs de Paris. Par MM G. Cuvier et Alex. Brongniart. Lu le 11 avril 1808. *Académie des sciences. Mémoires;* 1811; Année 1810(Première partie): 1–278 1 map 2 pl.

NOTE. Published in 1814. Read at the April 11, October 2 and October 16, 1808, sessions of the Académie. Expanded version of earlier publications, rec.234 and rec.241. Reissued as a monograph with preface, rec.691. See the latter record for more information on publication history.

285. Rapport, Fait à l'Institut, sur un Mémoire de M. Jacobson, intitulé: *Description anatomique d'un Organe observé dans les Mammifères. Paris. Muséum national d'histoire naturelle. Annales;* 1811; 18: 412–424.

NOTE. Report presented at the March 2, 1812 meeting of the Académie des sciences and signed by Tenon, Lacépède, and Cuvier. Also published in the *Journal de physique,* in 1812, v.75: 5–13, and in the *Académie des sciences, Procès verbaux;* 1914; 5(1812–1815): 23–27.

286. Sur les acères ou gastéropodes sans tentacules apparens; . . . (Extrait). *Société philomatique de Paris. Bulletin;* 1810, 1811; [Ser.II] 2(41): 221–222.

NOTE. Issue of Feb., 1811. Extract of rec.277.

287. Sur les os de Reptiles et de Poissons des carrières à plâtre [par M. Cuvier] et sur la structure du terrain des environs de Paris [par A. Brongniart]. *Société philomatique de Paris. Bulletin;* 1810, 1811; [Ser.II.] 2(41): 229–232.

NOTE. Issue dated February, 1811. Cuvier's part ends on p. 231 and is a summary of rec.278.

288. Rapport . . . sur des remarques de M. Férussac relatives à des *Terrains d'eau douce. Académie des sciences. Procès verbaux;* 1914; 5(1812–1815): 77–78.

NOTE. Reporter is not indicated. Read by Cuvier at the meeting of the Académie on July 27, 1812 and signed by Cuvier and Lamarck.

289. Recueil d'observations de zoologie et d'anatomie comparée, faites dans l'Océan-Atlantique, dans l'intérieur du Nouveau-Continent et dans la mer du Sud, . . . ; par Al. de Humboldt et A. Bonpland.—Premier volume. *Le Moniteur universel.* Paris; Dec.2, 1812; (337): 1332–1333.

NOTE. Review signed, G. Cuvier. The volume reviewed was published, Paris: Schoell; 1811; 368p.

290. Rapport Fait à la classe des Sciences mathématiques et physiques, sur divers Cétacés pris sur les côtes de France, principalement sur ceux qui sont échoués près de Paimpol, le 7 janvier 1812. *Paris. Muséum national d'histoire naturelle. Annales;* 1812; 19: 1–16. 1 pl.

NOTE. Presented at the April 6, 1812 meeting of the Académie des sciences and signed by Lacépède, Geoffroy-Saint Hilaire and Cuvier, as reporter. Published also in the *Académie des sciences, Procès verbaux;* 1914; 5 (1812–1815): 43–48, with the title "Rapport . . . sur les cétacés échoués de Saint Brieux" and with minor changes. An abstract was published, rec.293.

291. Sur un nouveau rapprochement à établir entre les classes qui composent le Règne animal. *Paris. Muséum national d'histoire naturelle. Annales;* 1812; 19: 73–84.

NOTE. Read at the May 25, 1812 meeting of the Académie des sciences. It was in this paper that Cuvier introduced his new concept of the division of the animal kingdom into four 'embranchements'. Translated into German in 1817 with editorial embellishments, rec.333. An extract was published in 1813, rec.299.

292. Sur la Composition de la Tête osseuse dans les Animaux vertébrés. *Paris. Muséum national d'histoire naturelle. Annales;* 1812; 19: 123–128.

NOTE. Read at the June 29 meeting of the Académie des sciences. Translated into German, rec.343. Extracted in French, rec.294, and also in German, rec.339.

293. Description des Cétacés échoués dans la baye de Paimpol. . . . (Extrait). *Société philomatique de Paris. Bulletin;* 1812, 1813; [Ser.II] 3(56): 69–71. 1 pl.

NOTE. Issue of May, 1812. Abstract of rec.290. Also published in *Le Moniteur universel,* Paris; 1812 June 5; (157): 614.

294. Sur la composition de la tête osseuse dans les animaux vertébrés. *Société philomatique de Paris. Bulletin;* 1812, 1813; [Ser.II] 3(59): [117]–119.

NOTE. Issue of Aug., 1812. Extract of rec.292.

295. [Nachricht über *fossile Schildkröten*]. *Taschenbuch für die gesammte Mineralogie,* Frankfurt; 1812; 6: 264–265.

NOTE. Slightly condensed translation of rec.279.

296. [Rapport sur] un Mémoire de M. Daudebert de Férussac intitulé: *Considérations générales sur les mollusques terrestres et fluviatiles et*

sur les terrains d'eau douce. Académie des sciences. Procès verbaux; 1914; 5(1812–1815): 160–161.

NOTE. Presented at the meeting of the Académie on Feb. 8, 1813 and signed by Lamarck, and Cuvier as reporter. The memoir reported on is by André É. J. F. Férussac, baron d'Audebard.

297. Rapport . . . sur un Mémoire de M. Marcel de Serres concernant les *Usages de différentes parties du tube intestinal dans les insectes. Académie des sciences. Procès verbaux;* 1914; 5(1812–1815): 185–186.

NOTE. Read at the meeting of the Académie on March 15, 1813, and signed by Lamarck, and Cuvier as reporter.

298. Note sur un poisson peu connu, pêché récemment dans le golfe de Gênes (le Lophote cépédien, *Giorna*). *Paris. Muséum national d'histoire naturelle. Annales;* 1813; 20: 393–400. 1 pl.

NOTE. Possibly read at the Académie des sciences on Nov. 2, 1813.

299. Sur un nouveau rapprochement à établir entre les classes qui composent le règne animal. *Société philomatique de Paris. Bulletin;* 1812, 1813; [Ser.II] 3(64): 201–202.

NOTE. Issue of Jan.,1813. Extract of rec.291.

300. Versuch einer mineralogischen Geographie der Gegend um Paris, von den HH. Cuvier, . . . , und Brongniart, . . . Frei ausgezogen von Gilbert. *Annalen der Physik,* Halle; 1813; 45(11): 229–290.

NOTE. Freely abstracted in German from the first three chapters of *Essai sur la géographie minéralogique* . . . (rec.691); cf. title and table of contents.

301. Observations *Sur le prétendu Homme témoin du Déluge* de Scheuchzer. *Journal des mines;* 1814; 36: 73–75.

NOTE. Another report on same paper was also published, rec.303. Source for above is given in article as *Gazette de Santé,* which was not located in the United States and was not seen.

302. Extrait D'une Note sur une tête presque entière de *palæeotherium,* retiré du gypse. *Journal des mines;* 1814; 36: 76–78.

NOTE. Read at the June 27, 1814 meeting of the Académie des sciences.

303. Nouvelles Observations sur le prétendu homme témoin du déluge de Scheuzer. *Société philomatique de Paris. Bulletin;* 1814: 22–23.

NOTE. Another report on same paper was also published, rec.301.

304. Mémoire sur la composition de la mâchoire supérieure des poissons, et sur le parti qu'on peut en tirer pour la distribution méthodique de ces animaux. *Société philomatique de Paris. Bulletin;* 1814: 73–78.

NOTE. Issue of Sept.,1814. Summary of paper published in full in 1815, rec.312.

305. Observations et recherches critiques sur différens Poissons de la Méditerranée et, à leur occasion, sur des Poissons des autres mers plus ou moins liés avec eux. *Société philomatique de Paris. Bulletin;* 1814: 80–92.

NOTE. Issue of Sept.,1814. Published in full in 1815, rec.313. The latter was translated in 3 sections into German, rec.347, rec.348, and rec.349.

306. Notice sur un poisson célèbre, et cependant presque inconnu des auteurs systématiques, appelé sur nos côtes de l'océan *Aigle* ou *Maigre,* sur celles de la Méditerranée, *Umbra, Fegaro* et *Poisson Royal,* avec une description abrégée de sa vessie natatoire. *Société philomatique de Paris. Bulletin;* 1814: 182–184.

NOTE. Full article published, rec.311. Not compared.

307. Rapport . . . sur les Mémoires de M. Savigny relatifs à *Certains animaux confondus jusqu'ici parmi les alcyons. Académie des sciences. Procès verbaux;* 1914; 5(1812–1815): 496–500.

NOTE. Presented at the meeting of the Académie on May 8, 1815 and signed by Lamarck, and Cuvier as reporter.

308. Rapport . . . sur le Mémoire de M. Savigny relatif aux *Organes de la bouche des insectes sans ailes. Académie des sciences. Procès verbaux;* 1914; 5(1812–1815): 521–526.

NOTE. Presented at the session of July 3, 1815 and signed by Lamarck, Cuvier, and Latreille with no indication of reporter.

309. Rapport . . . sur un Mémoire de M. Brongniart [sur les trilobites]. *Académie des sciences. Procès verbaux;* 1914; 5(1812–1815): 576–578.

NOTE. Presented at the meeting of Nov. 13, 1815 and signed by Lamarck, Ramond, and Cuvier as reporter.

310. [Sur le projet de loi relatif aux cours prévotales]. *Le Moniteur universel.* Paris; 1815 December 8; (342): 1361.

NOTE. Speech by Cuvier before a session of the Chambre des Députés, December 4, 1815.

311. Notice Sur un Poisson célèbre, et cependant presque inconnu des auteurs systématiques, appelé sur nos côtes de l'Océan, *Aigle* ou *Maigre,* et sur celles de la Méditerranée, *Umbra, Fegaro,* et *Poisson Royal:* avec une Description abrégée de sa vessie natatoire. Lue à la Classe des Sciences de l'Institut, en novembre 1813. *Paris. Muséum national d'histoire naturelle. Mémoires;* 1815; 1: 1–21. 3 pl.

NOTE. Read at the Académie des sciences, Nov. 8, 1813. A summary appeared in 1814, rec.306, and the full article, above, was translated into German in 1818, rec.346.

312. Mémoire Sur la composition de la Mâchoire supérieure des Poissons, et sur le parti que l'on peut en tirer pour la distribution méthodique de ces animaux. Lu à la Classe des Sciences de l'Institut, le 29 [i.e.28] mars 1814. *Paris. Muséum national d'histoire naturelle. Mémoires;* 1815; 1: 102–132.

NOTE. Extract was published in 1814, rec.304. Translated into German twice, rec.340 and rec.344. Corrected date of session comes from the *Académie des sciences, Procès verbaux;* 1914; 5(1812–1815): 338.

313. Observations et recherches critiques Sur différens Poissons de la Méditerranée, et à leur occasion sur des Poissons d'autres mers, plus ou moins liés avec eux. *Paris. Muséum national d'histoire naturelle. Mémoires;* 1815; 1: 226–241; 312–330; 353–363; 451–466. 3 pl.

NOTE. Read at the sessions of the Académie des sciences on May 23, May 30, and June 6, 1814. A summary appeared in 1814, rec.305. The full article, above, was translated into German in 1818 in three sections: rec.347, rec.348, and rec.349.

314. Mémoire sur les Ascidies et sur leur anatomie. *Paris. Muséum national d'histoire naturelle. Mémoires;* 1815; 2: 10–39. 3 pl.

NOTE. Read at the Oct. 17, 1814 session of the Académie des sciences. Published also in *Mémoires pour servir à l'histoire et à l'anatomie des mollusques,* (rec.724), no.20. An extract was published, rec.316, and a German translation, rec.370.

315. Mémoire sur les animaux des Anatifes et des Balanes Lam (*Lepas* Lin.) et sur leur anatomie. *Paris. Muséum national d'histoire naturelle. Mémoires;* 1815; 2: 85–101. 1 pl.

NOTE. Read at the May 22, 1815 session of the Académie des sciences. Also published in some issues of *Mémoires pour servir à l'histoire et à l'anatomie des mollusques,* (rec.724), no.23.

316. Mémoire sur les Ascidies et sur leur anatomie. *Société philomatique de Paris. Bulletin;* 1815: 10–14.

NOTE. Issue of Jan. 15. Extract of rec.314.

317. Extrait des Réflexions sur la marche actuelle des Sciences, et sur leurs Rapports avec la Société . . . (Séance de l'institut royal de France, le 24 avril 1816). *Annales maritimes et coloniales;* 1816; Ptie. 2, v.1: 250–251.

NOTE. Extracts from rec.318.

318. Réflexions sur la Marche actuelle des Sciences, et sur leurs Rapports avec la Société; . . . pour les sciences physiques; lues à la séance

publique d'installation le 24 avril 1816. *Le Moniteur universel.* Paris; 1816 May 22; (143): 589–590.

NOTE. Read at the public meeting of the four Academies on April 24, 1816. Also published in *Magasin encyclopédique,* 1816; 2: 313–338; in Cuvier's *Recueil des éloges historiques . . . ,* Paris, 1819–1827, and Brussels, Impression anastaltique, 1969; 1: 1–33; and in *Recueil des discours prononcés dans la séance publique annuelle de l'Institut royal de France Le mercredi 24 avril 1816,* Paris: Firmin Didot; 1816: 27–42. Translated into Italian, rec.321, and into Polish, rec.322. An extract was also published, rec.317.

319. [Discours de M. Cuvier, . . . Commissaire du roi pour la défense du projet de loi sur l'organisation des collèges électoraux, Prononcé à la Chambre des Députés, dans sa séance du 26 décembre 1816]. *Le Moniteur universel.* Paris; 1816 December 30; (365): 1462–1464.

NOTE. Published also separately, Paris, [1817?], Imprimerie de P. Gueffier. 43 p. 8°. Texts not compared.

320. Geognostische Betrachtungen, veranlasst durch Untersuchungen der fossilen Knochen vierfüssiger Thiere; von Cuvier, . . . (Ein beurtheilender Auszug, frei bearbeitet von Gilbert). *Annalen der Physik,* Halle; 1816; 52: 1–16 [i.e.p.117–132], 146–159 (p.133–145 are lacking in Library of Congress copy).

NOTE. Freely translated extract, including insertions by the translator, from the introduction (*Discours sur les révolutions de la surface du globe . . .*) to *Recherches sur les ossemens fossiles . . .* (rec.740) using also Jameson's English translation, rec.656 or rec.657; cf. p.1–3 of above.

321. Considerazioni sul corso attuale delle scienze e sulle loro relazioni colla società, del sig. cavaliere Cuvier, . . . lette nella pubblica sessione del 24 aprile 1816. *Biblioteca italiana,* Milan; 1816; 3(July/September): 147–156; 313–323.

NOTE. Translation of rec.318.

322. Rozwagi nad teraźnieyszym umiejętności kierunkiem, i nad jey stosunkami ze społecznością; Rzecz na posiedzeniu akademii umiejętności w Paryżu czytana przez P. Kiuwie (Cuvier)... [Reflections on the current trend in science and its relation to society. A speech delivered at a meeting of the Academy of Sciences in Paris by Chev. Cuvier, ...]. *Dziennik wilemski;* 1816 July/December; 4: 461–480.

NOTE. Translation by Roman K. Metelicki of rec.318. Not seen and not located out of Poland. The Library of the Polish Academy of Sciences, Institute of Zoology, Warsaw, has a copy.

323. [Rapport sur différents travaux de M. Moreau de Jonnès présentés à l'Académie, savoir: une carte géologique d'une partie de la Martinique; une mémoire sur la vipère jaune ou fer-de-lance de la Martinique et de Saint-Lucie; un recueil de règles d'hygiène applicables au climat des Antilles]. *Annales maritimes et coloniales;* 1817; Pt.3, 6: 151–153.

NOTE. Extracts from rec.616.

324. Mémoires de l'Académie impériale des sciences de Saint-Pétersbourg, tome IV, avec l'Histoire de l'Académie pour 1811;—tome V, avec l'Histoire de l'Académie pour 1812: 2 vol. in-4°. Pétersbourg, 1813 et 1815. Extrait de la partie physique. *Journal des savants;* 1817; (1842): 52–57.

NOTE. Comments by Cuvier on work of the Académie impériale des sciences, Saint-Pétersbourg (Akademiia nauk, St. Petersburg). The *Journal des savants* was originally published in 1817 and reprinted in 1842.

325. [Discours prononcé à la distribution des prix du Collège Royal de Louis-le-Grand, le 19 août 1817]. *Le Moniteur universel.* Paris; 1817 August 21; (233): 919–920.

NOTE. Published also separately, [n.p.; 1817?]; 6p. Texts not compared.

326. Rapport Sur un Mémoire de M. Dutrochet, . . . , intitulé *Recherches sur les Enveloppes du Foetus. Paris. Muséum national d'histoire naturelle. Mémoires;* 1817; 3: 82–97. 1 pl.

NOTE. Presented at the Académie des sciences meeting of Sept. 11, 1815. Signed by Percy, and Cuvier as reporter. Also published in *Académie des sciences, Procès verbaux;* 1914; 5(1812–1815): 547–552 which includes some material omitted from the above. Apparently also issued separately, 16 p. and pl., but not seen. Translated into German, rec.341.

327. Mémoire sur les oeufs des quadrupèdes. *Paris. Muséum national d'histoire naturelle. Mémoires;* 1817; 3: 98–119. 1 pl.

NOTE. Read at the Sept. 25, 1815 meeting of the Académie des sciences. Apparently also issued separately, 22 p. 4°, but not seen. Translated into German, rec.342, with many interspersed comments of editor. Also summarized in German, rec.359.

328. Extrait d'observations faites sur le cadavre d'une femme connue à Paris et à Londres sous le nom de Vénus Hottentote. *Paris. Muséum national d'histoire naturelle. Mémoires;* 1817; 3: 159[i.e.259]–274.

NOTE. Read at the Oct. 21, 1816 meeting of the Académie des sciences. Appeared also in various editions of *Discours sur les révolutions de la surface du globe . . .*, rec.679 etc. Translated into Italian with a few omissions, rec.566. An abridged and free German translation was published in 1819, rec.358.

329. Sur le genre Chironectes Cuv. (*Antennarius*. Commers.). *Paris. Muséum national d'histoire naturelle. Mémoires;* 1817; 3: 418–435. 3 pl.

330. Ueber die Würmer mit rothem Blute, von Cit. Cuvier. Geles. am 11. Nivos., an 10. *Isis von Oken,* Jena; 1817; 1: col.475–476.

NOTE. Read at the January 1, 1802 meeting of the Académie des sciences. Translation of rec.113. Another German translation was also published in 1803, rec.131.

331. Vom rothen Blute und dem Kreislaufe der Gliederwürmer. *Isis von Oken,* Jena; 1817; 1: col. 476–479.

NOTE. Republished from rec.719, a translation of *Leçons d'anatomie comparée,* v.4 p. 247–250, rec.712, but with occasional interspersed comments, presumably by Lorenz Oken.

332. Cuviers und Okens Zoologien neben einander gestellt, Okens Tabellen entworfen in der Abtey Kreuzlingen am Bodensee 1812, Cuviers im Nationalmuseum zu Paris 1816. *Isis von Oken,* Jena; [1817?]: col.1145–[1186].

NOTE. Translation of rec.291. Apparently a preprint of rec.333 which is v.1, 1817. Text is the same in both, but above lacks the introductory essay and remarks. The Bibliothèque nationale, Paris, has the only copy of the preprint found. Page numbers given on preprint are misnumbered or are omitted.

333. Cuviers und Okens Zoologien neben einander gestellt. *Isis von Oken,* Jena; 1817; 1: col.1145–[1200].

NOTE. Col.1145–1152 have title, "Ueber die Einführung einer neuen Zusammenrückung der Klassen, woraus das Thierreich besteht, von Cuvier," which is a translation of rec.291 with editorial comments inserted and with some omissions which mention works of other scientists. What appears to be a preprint, rec.332, was also published, but without the opening remarks and introductory essay.

334. O rozwijaniu się fizycznem i moralnem człowieka. Wyjątek przetłomaczony z dzieła . . . pod tytułem Le règne animal distribué d'après son organisation / przez/ S. [On the physical and moral development of man. *Pamietnik Warszawski czyli Dziennik Nauk i Umiejętności;* 1817; An. 3(8): 467–481.

NOTE. Translation from *Le Règne animal,* 1817, rec.747, v.1: 75–84; 1829, rec.748, v.1: 75–80. Not seen.

335. Discours [prononcés dans la séance publique tenue par l'Académie française, Pour la réception de M. Cuvier, le 27 août 1818.]. *Le Moniteur universel,* Paris; 1818 September 7: 1063–1064.

NOTE. Also published separately, Paris, Firmin Didot, Imprimeur, 1818 20 p. 8° and in Cuvier's *Recueil des éloges historiques . . . ,*

Paris, 1819–1827, v.2, p.443–467, and Impression anastaltique, Bruxelles, 1969.

336. [Tête d'Orang-outang]. *Nouveau journal de médecine;* 1818; 1: 97–98.

NOTE. Académie des sciences, Séance du 16 février.

337. Sur les Diodons vulgairement Orbes-épineux. *Paris. Muséum national d'histoire naturelle. Mémoires;* 1818; 4: 121–138. 2 pl.

338. Sur les poissons du sous-genre *Mylètes. Paris. Muséum national d'histoire naturelle. Mémoires;* 1818; 4: 444–456. 2 pl.

339. Cuvier über die Zusammensetzung des knöchernen Kopfes der Wirbelthiere. *Deutsches Archiv für die Physiologie,* Halle; 1818; 4: 244–247.

NOTE. German extracts from rec.292.

340. G. Cuvier über die Zusammensetzung des Oberkiefers der Fische, und die Benutzung derselben zur methodischen Eintheilung dieser Thiere. *Deutsches Archiv für die Physiologie,* Halle; 1818; 4: 247–262.

NOTE. Translation of rec.312, but a different translation from rec.344.

341. Ueber die Abhandlung von Dutrochet Arzt zu Chateau-Renaud: betitelt: Untersuchungen über die Fötushüllen v. G. Cuvier.... *Isis von Oken,* Jena; 1818; 2/3: col.114–126.

NOTE. Translation of rec.326, interspersed with numerous editorial comments etc.

342. Ueber die Eyer der Vierfüssler, v. Cuvier. . . . *Isis von Oken,* Jena; 1818; 2/3: col.126–139. 1 pl.

NOTE. Translation of rec.327. Not the same translation as rec.359.

343. Ueber die Zusammensetzung des Schädels bey den Wirbelthieren von Cuvier. *Isis von Oken,* Jena; 1818; 2/3: col.283–285.

NOTE. Translation of rec.292, with condensation of one section.

344. Ueber die Zusammensetzung des Oberkiefers der Fische, und die Anwendung, welche man davon auf die methodische Eintheilung dieser Thiere machen kann von Cuvier. *Isis von Oken,* Jena; 1818; 2/3: col.285–289; col.1059–1072.

NOTE. Translation of rec.312. Another German translation was also published, rec.340.

345. Osteologie der Crocodille v. Cuvier. . . . *Isis von Oken,* Jena; 1818; 2/3: col.289–292.

NOTE. German condensation of rec.244 consisting only of introduction and sections on skull and lower jaw.

346. Ueber Sciaena Umbra, v. Cuvier, Bemerkungen über einen berühmten und dem ohnerachtet bey den Systematikern beynah unbekannten Fisch, genannt auf unsern Küsten des Weltmeers Aigle oder Maigre, und auf denen des mittelländischen Meers Umbra, Fegaro und Poisson royal; mit einer kurzen Beschreibung seiner Schwimmblase. . . . *Isis von Oken,* Jena; 1818; 2/3: col.1170–1182. 1 pl.

NOTE. Translation of rec.311.

347. Ueber Argentina Sphyraena, Mulla imberbis, von G. Cuvier. *Isis von Oken,* Jena; 1818; 2/3: col.1182–1191. 1 pl.

NOTE. Partial translation of rec.313. Translation continued by rec.348 and rec.349.

348. Ueber Ophidium imberbe (Donzelle), Coryphaena Novacula (Rason), Girelle Coryphaena, aus dem Mittelmeer, und einige Verwandte v. Cuvier. *Isis von Oken,* Jena; 1818; 2/3: col.1636–1646. 1 pl.

NOTE. Partial translation of rec.313. Above translation is preceded by rec.347 and completed by rec.349.

349. Fortsetzung der kritischen Beobachtungen und Untersuchungen über verschiedene Fische des Mittelmeeres, und gelegentlich über Fische anderer Meere, die mehr oder weniger mit jenen in Verbindung stehen, von Mr. G. Cuvier. [*Labrus, Chromis, Crenilabrus, Lutjanus, Anthias*]. *Isis von Oken,* Jena; 1818; 2/3: col.1656–1670. 1 pl.

NOTE. With rec.347 and rec.348, completes translation of rec.313.

350. Discours . . . Sur l'Évaluation des Produits indirects et variables. Séance du 30 juin 1819. [Paris]: Hacquart, imprimeur; [1819?]. 18 p. 8°. (France. Chambre des députés. Impressions ordonnées. Session 1818; v.4. no. 180).

NOTE. At head of title: Chambre des Députés.

351. Cuvier, Frédéric, [?] 1773–1838. [Review of Geoffroy Saint Hilaire's] *Philosophie anatomique; des organes respirateurs, sous le rapport de la détermination et de l'identité de leurs pièces osseuses. Journal complémentaire du Dictionnaire des sciences médicales;* 1819; 3: 142–157.

NOTE. Although this article is signed "G. Cuvier," the journal index attributes it to F. Cuvier and the text refers to G. Cuvier in the third person throughout.

352. [Discours sur la liberté de la presse. Séance du 15 avril, Chambre des Députés]. *Le Moniteur universel.* Paris; 1819 April 16; (106): 455–456.

353. [Discours sur la liberté de la presse. Séance du 17 avril, Chambre des Députés]. *Le Moniteur universel.* Paris; 1819 April 19; (109): 470.

354. [Discours sur le projet relatif à la poursuite des délits de la presse. Séance du 20 avril, Chambre des Députés]. *Le Moniteur universel.* Paris; 1819 April 22; (112): 485–486.

355. [Discussion sur le projet de la loi relatif aux dépenses. Séance du 27 mai, Chambre des Députés]. *Le Moniteur universel.* Paris; 1819 May 29; (149): 685–686.

356. [Discours sur les voies et moyens. Chambre des Députés, Séance du 30 juin]. *Le Moniteur universel.* Paris; 1819 July 2; (183): 832–833.

357. Sur les poissons du sous-genre *Hydrocyn,* sur deux nouvelles espèces de *Chalceus,* sur trois nouvelles espèces de *Serrasalmes,* et sur l'*Argentina glossodonta* de Forskahl, qui est l'*Albula gonorhynchus* de Bloch. *Paris. Muséum national d'histoire naturelle. Mémoires;* 1819; 5: 351–379. 3 pl.

NOTE. Summarized in rec.362.

358. G. Cuvier Bemerkungen an der Leiche einer Hottentottin. . . . *Deutsches Archiv für die Physiologie,* Halle; 1819; 5: 153–159.

NOTE. Abridged and free translation of rec.328.

359. G. Cuvier über das Ei der Säugethiere. . . . *Deutsches Archiv für die Physiologie,* Halle; 1819; 5: 574–584.

NOTE. Shortened translation of rec.327. Not the same translation as rec.342.

360. Ueber die Schüssel-Schnecken und ihre Verwandten; von Cuvier (Mollusques). *Isis von Oken,* Jena; 1819; 4/5: col.723–742. 2 pl.

NOTE. Translation of Article xviii, *Mémoire sur l'Haliotide,* . . . of *Mémoires pour servir à l'histoire et à l'anatomie des mollusques,* (rec.724).

361. Extraits des analises des travaux de l'Académie royale des sciences, pendant les années 1819 et 1820 en ce qui concerne l'anatomie comparative du cerveau, et les lois de l'ostéogénie. [Paris: Imprimerie de Rignoux; 182–?]. 34p. 8°.

NOTE. Cover title. On verso of cover title: "Extraits des rapports présentés à l'Académie . . . , en ce qui concerne les travaux de M. le docteur Serres, . . . " Extracts from rec.623 and rec.625. Also published in rec.433 without "Les lois de l'ostéogénie." Translated into German and slightly condensed, rec.388.

362. Sur les poissons du sous genre *Hydrocyon;* sur deux espèces nouvelles de *Chalceus;* sur trois espèces nouvelles de *Serrasalmes* et sur l'argentin *Glossodonte* de Forskahl qui est l'*Albula Gonorhynchus* de

Bloch. *Annales générales des sciences physiques,* Bruxelles; 1820; 4: 60–62.

NOTE. Summary of rec.357.

363. Sur une nouvelle espèce de Dindon de la baie de Honduras. *Meleagris ocellata. Annales générales des sciences physiques,* Bruxelles; 1820; 7: 145–146.

NOTE. Summary of rec.368. Above was translated into Italian, rec.382.

364. [Rapport fait à l'Académie des sciences de Paris, sur un ouvrage de M. J. Victor Audouin, ayant pour titre: *Recherches anatomiques sur le thorax des animaux articulés, et celui des insectes en particulier.*] Séance du 19 février. *Annales générales des sciences physiques*, Bruxelles; 1820 [pub., 1821?]; 7: 396–406. 1 pl.

NOTE. Report presented at the Feb. 19, 1821 meeting of the Académie des sciences and signed by Lacépède, Duméril, and Cuvier as reporter. This was published separately; [n.p.; n.d.] 11p. 1pl. and again; Paris, Imprimerie de J. Tastu, 1823, 15 p., 4°. It was published in *Académie des sciences, Procès verbaux;* 1916; 7(1820–1823): 144–147, and also in French in *Isis von Oken,* Jena, 10/11, 1822, col. 80–87. A number of copies of the 11 page separate have identical manuscript directions for a rearrangement of text which is corrected in all the other printings except the *Annales* . . . and the 11 page separate itself. The corrected version was translated into German, rec.389. A truncated version of the corrected printing was also published, rec.374.

365. [Discours . . . Sur le Projet de Loi relatif aux Élections; prononcé dans la séance du 24 mai 1820]. *Le Moniteur universel.* Paris; 1820 May 25: 713–717.

NOTE. Also published separately, [Paris, Imprimerie de Mme. veuve Agasse], 1820 30p. 8°. Title above is from separate.

366. [Discours prononcé . . . sur l'Article 14 de la loi des recettes, instruction publique. Chambre des Députés. Séance du 6 juillet 1820]. *Le Moniteur universel.* Paris; 1820 July 8; (190): 973–974.

NOTE. Published separately: [Paris: Hacquart, Imprimeur; 1820?]; 21p. 8°. Texts not compared.

367. Rapport-verbal fait à l'Académie des sciences . . . sur l'ouvrage de M. de Férussac. *Le Moniteur universel.* Paris; 1820 August 1; (214): 1089–1090.

NOTE. Report made to the Académie des sciences, March 6, 1820. Also published in *Isis von Oken,* Jena, 1822 10/11 (*Litterarischer Anzeiger*): 116–119, in French with title: *Sur l'Ouvrage intitulé: Histoire naturelle, générale et particulière des mollusques [terrestres] et fluviatiles* . . . Par le Baron de Férussac.

368. Description d'une nouvelle espèce de Dindon de la Baie de Honduras . . . (*Meleagris ocellata,* Cuv.). *Paris. Muséum national d'histoire naturelle. Mémoires;* 1820; 6: 1–4. 1 pl.

NOTE. A summary, rec.363, was translated into Italian, rec.382.

369. Discours prononcé par M. le Baron Cuvier exerçant les fonctions de Président du Conseil Royal de l'Instruction publique. Paris. Université, Faculté de Médicine. Séance publique du 7 novembre 1820; [1821?]: 15–16.

NOTE. Possibly same text as rec.375, but not compared.

370. Abhandlung über die Ascidiae u. über ihre Anatomie. *Isis von Oken,* Jena; 1820; 6/7: 387–404. 2 pl.

NOTE. Translation of rec.314.

371. Ueber *Thalia* Brown und *Salpa* Forskal. . . . *Isis von Oken,* Jena; 1820; 6/7(*Literarischen Anzeiger*): 260–272. 1 pl.

NOTE. Translation of rec.148.

372. Prospectus, par M. le baron Cuvier. In: Temminck, C. J. Nouveau recueil de planches coloriées d'oiseaux, pour servir de suite et de complément aux planches enluminées de Buffon, . . . ; pub. par C. J. Temminck, . . . et Le Baron Meiffren Laugier de Chartreuse, . . . Paris

and Strasbourg: F. G. Levrault; Amsterdam: Legras Imbert; [1820?]: v.1, p.[1]–8 folio.

NOTE. A. Belin, pr. Published also in 4°, v.1, p.1–11 and apparently separately. Published in French in *Isis von Oken,* Jena, 1823 12 Hft.5 (Suppl.) 2, 1823, without page or column numbers, but not compared with above. Other printings and editions of Temminck were not seen.

373. Discours . . . , Relatif à l'allocation de 50,000 fr. pour encouragemens à l'instruction primaire. Séance du 12 Juin 1821. [Paris: Hacquart; 1821?]. 16p. 8°. (France. Chambre des Députés, 1814–1848. Impressions ordonnées. Session 1820; v. IV no.147).

NOTE. At head of title: Chambre des Députés.

374. Extrait d' un rapport de M. Cuvier, fait à l'Académie des Sciences, dans la séance du 19 février, sur un ouvrage de M. Audouin, ayant pour titre: *Recherches anatomiques sur le thorax des animaux articulés, et celui des insectes hexapèdes en particulier. Journal de physiologie expérimentale et pathologique;* 1821; 1(2): 152–164. 1 pl.

NOTE. A slightly shortened printing of the corrected version of rec.364.

375. [Hommage à la médecine], réponse à M. Richerand. Séance publique de la Faculté de médecine de Paris du 7 novembre 1820. *Le Moniteur universel.* Paris; 1821 January 2; (2): 7.

NOTE. Possibly same text as rec.369, but not compared.

376. [Discours sur le projet de loi relatif au remboursement du premier cinquième des reconnaissances de liquidation. Chambre des Députés, Séance du 16 février 1821]. *Le Moniteur universel.* Paris; 1821 February 18; (49): 211–212.

NOTE. Also published separately: [Paris, Hacquart; 1821?]; 19 p. (France. Chambre des Députés. Impressions ordonnées. Session de 1820. T.I no.35). Cuvier spoke again on this law, rec.377.

377. [Discours sur le projet de loi . . . relatif au remboursement du premier cinquième des reconnaissances de liquidation]. Chambre des Pairs, Séance du 7 mars. *Le Moniteur universel.* Paris; 1821 March 16; (75): 353.

NOTE. Cuvier also spoke on this law at the session of February 16, 1821, rec.376.

378. [Discours sur le projet de la loi des finances. (Budget du conseil royal de l'instruction publique). Chambre des députés, Séance du 18 juin]. *Le Moniteur universel.* Paris; 1821 June 20; (171): 894.

NOTE. Also published separately, [Paris, Hacquart; 1821?] 14 p. (France. Chambre des Députés, 1814–1848. Impressions ordonnées. Session 1820; IV. no.182).

379. [Réponse sur la distribution des prix aux élèves des collèges royaux de Versailles . . . par le conseil royal de l'instruction publique, 16 août 1821]. *Le Moniteur universel.* Paris; 1821 August 17; (228 and 229): 1199.

380. [Défense du projet de loi relatif au régime sanitaire, Chambre des Pairs, Séances du 11, 13 & 14 décembre, 1821]. *Le Moniteur universel.* Paris; 1821 December 19; (353): 1701–1702; 1821 December 24; (358): 1726; 1821 December 28; (362): 1740.

381. Discours prononcé par M. le Baron Cuvier, . . . , exerçant les fonctions de Président du Conseil Royal de l'Instruction publique. Paris. Université. Faculté de Médecine. Séance publique; Nov. 22, 1821: 49–52.

382. Nuova specie di pollo d'India della Baja di Honduras. *Meleagris ocellata.* Del. Sig. Cuvier. *Giornale di fisica, chimica, storia naturale, medicina ed arti,* Pavia; 1821; Ser.II 4: 164.

NOTE. Translation of rec.363.

383. Mémoire sur les Ossemens de fossiles analogues aux Tapirs. *Bibliothèque universelle des sciences, belles-lettres, et arts. Sciences et arts,* Genève; 1822; 19: 294–296.

NOTE. Paper read by Cuvier at the August 13, 1821 meeting of the Académie des sciences. A condensed version was published in *Recherches sur les ossemens fossiles* . . . 1821–1824, rec.742, v.2.

384. Discours . . . Pour la défense du projet de loi relatif à la répression des délits commis par la voie de la presse, ou par tout autre moyen de publication. Imprimé par ordre de la Chambre. [Paris: Didot l'aîné; 1822]. 27p. 8°. (France. Chambre des Pairs. Impressions diverses. Session de 1821; v. 2, no.50).

NOTE. At head of title: Chambre des Pairs de France. Session de 1821. Séance du vendredi Ier. mars 1822.

385. [Éclaircissemens sur le budget du conseil royal de l'instruction publique. Chambre des Députés, Séance du 27 mars]. *Le Moniteur universel.* Paris; 1822 March 29; (88): 480–481.

NOTE. Also published separately, [Paris: Hacquart; 1822]; 14p. 21.5 cm. (France. Chambre des Députés. Session 1821 [i.e.1822], Impressions ordonnées, T.II. Nº139).

386. Observations . . . , Sur le Budget du Conseil royal de l'Instruction publique. Imprimées par ordre de la Chambre. Séance du 27 Juillet 1822. [Paris: Hacquart; 1822?]. 6p. 8°. (France. Chambre des Députés. Impressions ordonnées . . . Session de 1822; v. IV, no.201).

NOTE. At head of title: Chambre des Députés. Session de 1822.

387. Rapport sur un mémoire de M. Flourens, intitulé: *Détermination des propriétés du Système nerveux, ou Recherches physiques sur l'irritabilité et la Sensibilité. Paris. Muséum national d'histoire naturelle. Mémoires;* 1822; 9: 120–138.

NOTE. Presented at the meeting of the Académie des sciences, July 22, 1822. Report with MM. Portal, le Cte. Berthollet, Pinel, Duméril and Cuvier as reporter. Published also in the *Revue encyclopédique,* 16 (47): 249[i.e.229]–246, 1822; *Journal de physiologie expérimentale et pathologique,* 2 (4): 371–384, 1822; *Isis von Oken,* Jena [12/13], col. 514–523, 1823, (in French); *Académie des sciences, Procès verbaux;* 1916; 7(1820–1823): 349–354; *Journal complé-*

mentaire des sciences médicales, 15: 44–57, 1823; and appeared separately, n.p., n.d., 18p. Translated into English, rec.392.

388. Serres über die vergleichende Anatomie des Gehirns. Paris 1821 (Bericht von G. Cuvier vor den öffentlichen Sitzung der Akademie der Wissenschaften zu Paris am 2ten April 1821.). *Deutsches Archiv für die Physiologie,* Halle; 1822; 7: 337–349.

NOTE. Slightly condensed translation of rec.361 without the section on the 'lois de l'ostéogénie', and of rec.433.

389. Cuvier's Bericht über J. V. Audouin's anatomische Untersuchungen über den Thorax der Gliederthiere überhaupt und der Insekten insbesondere. . . *Deutsches Archiv für die Physiologie,* Halle; 1822; 7: 435–444.

NOTE. Translation of corrected version of rec.364.

390. Observations . . . , Sur le Budget du Ministère de l'Intérieur (Instruction publique), Séance du 7 avril 1823. [Paris: Hacquart; 1823?]. 7p. 8°. (France. Chambre des Députés. Impressions ordonnées. Session de 1823; v. I. no.59).

NOTE. At head of title: Chambre des Députés. Session de 1823.

391. Discours . . . , Sur le Budget du Conseil royal de l'instruction publique. . . . Séance du 10 avril 1823. [Paris: Hacquart; 1823?]. 6p. 8°. (France. Chambre des Députés. Impressions ordonnées. Session de 1823; v. I. no.96).

NOTE. At head of title: Chambre des députés. Session de 1823.

392. Report made to the Academy of Sciences on a Paper of M. Flourens, entitled *Determination of the properties of the Nervous System, or Physical Researches concerning Irritability and Sensibility.* By M. G. Cuvier. *Philosophical magazine,* London; 1823; 61: 114–126.

NOTE. Translation of rec.387.

393. Über eine neue urweltliche, zur Klasse der dickhäutigen Thiere gehörende Thiergattung, *Anthracotherium* oder Kohlenthier. *Archiv für die neuesten Entdeckungen aus der Urwelt,* Leipzig; 1823; 5: 122–136.

NOTE. Translation of p.396–405, v.3, 1822, of *Recherches sur les ossemens fossiles,* rec.742. Probably the paper read at the Académie des sciences on June 10, 1822.

394. Über die knochenhaltigen Kalk-Breccien des mitteländischen Meerbeckens. Abgekürtze Uebersetzung des 4ten Kapitels im 4ten Theil der Recherches sur les Ossemens fossiles par M. le Baron G. Cuvier. Paris 1823. . . . *Archiv für die neuesten Entdeckungen aus der Urwelt,* Leipzig; 1823; 5: 211–269.

NOTE. Abridged translation of p. 167–226, v.4, *Recherches sur les ossemens fossiles,* (rec.742); cf. title. See also rec.255.

395. Sur la détermination des diverses espèces de baleines vivantes. . . . (Recherches sur les Ossemens Fossiles, tome V). *Annales des sciences naturelles;* 1824; 2: 27–41.

NOTE. Published also on p.359–370 of *Recherches sur les ossemens fossiles,* v.5, 1821–1824, rec.742. Plate mentioned in text was not found in copies seen and may refer to that in the *Recherches . . . .*

396. Sur des os de Sèche fossiles. *Annales des sciences naturelles;* 1824; 2: 482–485. 1 pl.

397. Résumé des observations . . . sur les espèces de Crocodiles fossiles. *Bulletin des sciences naturelles;* 1824; 1: 93–96.

NOTE. Translated into German, rec.403.

398. Remarques sur de prétendus fossiles humains. . . . [Ext. du compte rendu des travaux de l'Acad. des Sciences, pour l'année 1823]. *Bulletin des sciences naturelles;* 1824; 2: 280–281.

NOTE. Extracts from rec.631. Published much more fully in rec.401.

399. Description géognastique des environs du Puy en Velay . . . par M. J. M. Bertrand-Roux [Bertrand de Doue]. [Rapport]. *Bulletin des sciences naturelles;* 1824; 3: 1–3.

NOTE. Extract from rec.631.

400. [Discours . . . , Sur le Chapitre V du Budget du Ministre de l'intérieur. Séance du 14 juillet 1824]. *Le Moniteur universel.* Paris; 1824 July 16; (198): 985.

NOTE. Published also separately, Paris, Hacquart, 7 p. 8° (France. Chambre des Députés. Session de 1824. [Publication] no. 274).

401. Nouvelles observations Sur une altération singulière de quelque Têtes humaines, Lues à l'Académie des Sciences, le 16 juin 1823. *Paris. Muséum national d'histoire naturelle. Mémoires;* 1824; 11: 321–334.

NOTE. Translated into German, rec.410. A brief extract of the above was also published, rec.398.

402. Extrait d'un rapport sur l'état de l'Histoire naturelle et sur ses accroissements depuis le retour de la paix maritime; . . . In: Recueil des discours prononcés dans la séance publique annuelle de l'Institut Royal de France, le samedi 24 avril 1824. Paris: Imprimerie de Firmin Didot; 1824: 35–53. 4°.

NOTE. Cuvier's reprint or preprint from above in the Library of the Muséum national d'histoire naturelle, Paris, has its own covering title page. Pages in both are misnumbered, with no numbers, 37–38. Also published in the *Rapport historique sur les progrès des sciences naturelles. . .* , rec.694; and in Cuvier's *Recueil des éloges historiques. . .* , Paris, 1819–1827, and Brussels, Impression anastaltique, 1969, v.3, 449–478. Translated into English, rec.476.

403. Cuvier's allgemeine Übersicht des Vorkommens verschiedener, bis jetzt entdeckter urweltlicher Krokodil-Arten. *Archiv für die neuesten Entdeckungen aus der Urwelt,* Leipzig; 1824; 6: 127–130.

NOTE. Translation of rec.397.

404. Rapport . . . sur les nouveaux Mémoires de M. Flourens sur le *Cerveau des poissons,* sur la *Régénération des parties du crâne et de l'encéphale,* et sur les *Fonctions propres aux diverses parties de l'oreille. Académie des sciences. Procès verbaux;* 1918; 8(1824–1827): 229–231.

NOTE. Presented at the June 13, 1825 meeting of the Académie and signed by Humboldt, Portal, Duméril, Dulong, and Cuvier as reporter.

405. Rapport sur la partie zoologique de l'Expédition Duperrey . . . Fait à l'Académie des Sciences, séance du [18] juillet 1825. *Annales des sciences naturelles;* 1825; 6: 5–20.

NOTE. Expedition on the corvette, S. M. La Coquille. Presented at the July 18, 1825 meeting of the Académie des sciences by Cuvier and signed by him as reporter and by Latreille. Published also in the *Annales maritimes et coloniales;* 1825 [26],Pt.2: 189–202; separately, [Paris, Imprimerie royale, September, 1825] 16p. 8°; in R. P. Lesson, *Journal d'un voyage pittoresque autour du monde. . .* Paris, Amable Gobin, 1830, p.xxviij–xlij; and in the *Académie des sciences, Procès verbaux;* 1918; 8(1824–1827): 241–246. A very condensed report was also published, rec.407; and a German abstract was published in 1831, rec.534.

406. Rapport sur les progrès des sciences physiques pendant l'année 1824. (1er extrait) Météorologie.—Sur les défrichements de forêts depuis 1819; et sur leur influences. *Le Globe.* Paris; 1825 July 2; (127): 654.

NOTE. Brief extract from the Météorologie section of Cuvier's annual "Analyse des travaux. . . " for the year 1824 which was published in 1827, rec.634.

407. [Rapport sur la partie zoologique de l'expédition sous les ordres de M. Duperré]. *Le Globe.* Paris; 1825 July 21; (135): 694.

NOTE. Presented at the July 18, 1825 meeting of the Académie des sciences. A very brief condensation of Cuvier's report as published in rec.405.

408. Duméril, André Marie Constant, 1774–1860. Le rapport de M. le baron Cuvier et de M. le professeur Duméril, . . . In: Barry, Sir David, 1780–1835. Recherches expérimentales sur les causes du mouvement du sang dans les veines; Mémoire lu à l'académie des sciences le 8 juin 1825, par David Barry, . . . Avec le Rapport . . . Paris: Crevot et chez l'auteur; 1825: 64–74.

NOTE. Didot le jeune, pr. Although Duméril is the reporter and Cuvier the other signer of this report, it is included here because it appears under Cuvier as author in some catalogues and bibliographies. Presented at the meeting of the Académie, on August 29, 1825. Also published in French with English title in the English edition of the above: Barry, Sir David. *Experimental researches on the influence exercised by atmospheric pressure . . .* , London, Underwood, 1826: 166–173.

409. On the Osteology of Reptiles, and on the Geological Position of their Fossil Remains. By M. Le Baron G. Cuvier. *Philosophical magazine,* London; 1825; 65: 447–457.

NOTE. Translation of *Recherches sur les ossemens fossiles,* (rec.743), v.5, part 2, 1825, p.1–12.

410. Neue Beobachtungen über eine merkwürdige Veränderung einiger menschlichen Köpfe. Vom Baron Cuvier. *Notizen aus dem Gebiete der Natur= und Heilkunde,* Erfurt; 1825; 9: 337–343.

NOTE. Translation of rec.401.

411. Rapport . . . sur le Mémoire de MM. Quoy et Gaymard relatif aux *Mollusques et aux Zoophytes qu'ils ont observés dans la baie d'Algésiras. Académie des sciences. Procès verbaux;* 1918; 8(1824–1827): 435–438.

NOTE. Report presented at the October 2, 1826 meeting of the Académie and signed by Latreille, and by Cuvier as reporter. Published with a substantial omission in 1827, rec.423.

412. Rapport verbal sur un ouvrage intitulé: *Recherches sur les ossemens fossiles du département du Puy-de-Dôme;* (Fait à l'Acad. royale

des Sciences, le 23 octobre 1825 [i.e.1826]). *Annales des sciences naturelles;* 1826; 9: 273–278.

NOTE. Report on memoir by Bravard, Croizet, and Jobert. Published also separately; [Paris, Crochard, 1826]. 6p. 8°, and in *Le Globe,* rec.415, which was was not compared with above. Continued by rec.441.

413. [Éclaircissements relativement au crocodile de M. Caillaud]. *Le Globe*. Paris; 1826 February 11; 3(22): 119.

NOTE. Verbal response of Cuvier, at a meeting of the Académie des sciences, to E. Geoffroy St. Hilaire's paper, *Sur les crocodiles d'Égypte*.

414. Analyse du discours de M. Cuvier sur les progrès récents de la chimie [prononcé dans la séance générale de l'Institut]. *Le Globe*. Paris; 1826 May 13 and May 16; (61): 328 and (62): 330–331.

NOTE. The second part in issue 62, has the same title as above with the substitution as part of the title, "Prononcé dans la séance publique des quatre académies." Condensation of rec.419 or of the address itself. Not same condensation as rec.418. Translated into German, rec.420, with omissions and slight differences.

415. Ossements fossiles de Boulade en Auvergne. *Le Globe*. Paris; 1826 October 28; 4(33): 175.

NOTE. Verbal report by Cuvier given at the October 23, 1826 meeting of the Académie des sciences on the " . . . ossements fossiles trouvés dans le Département du Puy-de-Dôme, et publiée par MM. Auguste Bravart, . . . , l'abbé Croisé et Jobert aîné." Another publication, rec.412, of the same report was not compared with the above.

416. [Mémoire sur le genre de reptiles batraciens nommé Amphiuma, et sur une nouvelle espèce de ce genre: *Amphiuma tridactylum*]. *Le Globe*. Paris; 1826 November 16; 4(41): 216.

NOTE. Read at the Académie des sciences, November 13, 1826. Published also in *Bulletin des sciences naturelles,* 1827; 10: 162–163. Published more fully, rec.429.

417. [Discours pour la défense du projet de loi relatif aux écoles secondaires de médecine. Chambre des Pairs. Séance du mardi 2 mai]. *Le Moniteur universel.* Paris; 1826 May 3; (123): 643–644.

NOTE. Also published separately, [n.p.; 1826]; 15p., 21 cm. (France. Chambre des Pairs, 1814–18. Impressions ordonnées. III. Nº 94, 1826) with occasional slight differences in wording and with omissions.

418. [. . . extrait d'un rapport sur les *changemens éprouvés par la théorie chimique, et sur quelques nouvelles applications de la chimie aux besoins de la société*]. *Revue médicale française et étrangère;* 1826; 2: 331–334.

NOTE. Extracts and condensations of rec.419, but not same condensation as rec.414. Presented at the Séance publique annuelle des quatre académies on April 24, [1826].

419. Extrait d'un rapport sur les principaux changements éprouvés par les théories chimiques et sur une partie des nouveaux services rendus par la chimie à la société. In: Institut de France. Séance publique annuelle des quatre Académies, du lundi 24 avril 1826. Paris: Imprimerie de Firmin Didot; 1826: 12–29. 4°.

NOTE. Also published separately, [Paris, Imprimerie de Firmin Didot, n.d.], 18p. 4°; and in Cuvier's *Recueil des éloges historiques. . .* Paris, 1819–1827 and Bruxelles, Impression anastaltique, 1969, v.3, p.479–506. Rec.414 and rec.418 are condensations. Translated and condensed in German, rec.420.

420. Ueber die neuesten Fortschritte der Chemie. *Notizen aus dem Gebiete der Natur= und Heilkunde,* Erfurt; 1826; 14: col.209–214.

NOTE. Translation of rec.414 with some omissions, and with inclusions from rec.419.

421. O skłonności zwierząt do życia w towarzystwie. [On tendency of animals to live in association with other animals]. *Biblioteka Polska;* 1826; 2: 97–110, 155–167.

NOTE. Not seen.

422. Rapport fait à l'Académie . . . sur les ossemens de la grotte d'Oiselles, [qui ont été addressés par le Préfet du Doubs à son Ex. M. le Ministre de l'Intérieur]. *Annales de chimie et de physique;* 1827; Ser.II.36: 208–216.

NOTE. Presented at the July 16, 1827 session of the Académie des sciences, by Cuvier. Signed by Brongniart, Beudant, and Cuvier as reporter. Published also in the *Académie des sciences, Procès verbaux;* 1918; 8(1824–1827): 568–570. A condensed report was also published, rec.427, and translated into English, rec.434. What apparently is a Russian translation and condensation, rec.455, was not seen.

423. Extrait du Rapport sur les Observations zoologiques de MM. Quoy et Gaimard. Fait à l'Académie des Sciences, séance du 2 octobre 1826. *Annales des sciences naturelles;* 1827; 10: 435–438.

NOTE. Report signed by Latreille, and Cuvier as reporter. Published more fully in 1918, rec.411.

424. Histoire naturelle des Poissons. *Annales des sciences naturelles;* 1827; 12: 396–414.

NOTE. "Ces considérations préliminaires sont extraites du Prospectus de l'ouvrage par M. le baron Cuvier," footnote #2, p.396. Same as rec.705 with omission of section on authors' indebtedness to individual travelers. Translated into German, rec.453.

425. Sur le genre de reptiles batraciens nommé Amphiuma, et sur une nouvelle espèce de ce genre:—*Amphiuma tridactylum*, . . . *Bulletin des sciences naturelles;* 1827; 12: 132–134.

NOTE. Condensation of rec.429.

426. [Mémoire sur le Scare; . . . Académie des Sciences. Séance du 25 juin 1827]. *Le Globe*. Paris; 1827 June 28; 5(27): 196.

NOTE. Also published in *Bulletin des sciences naturelles;* 1827; 11: 444–445. Translated into English, rec.435.

427. Caverne à ossements.—Grotte d'Oiselles, près Besançon. *Le Globe*. Paris; 1827 July 19; 5(46): 243–244.

NOTE. At head of column: France. Académie des Sciences Séance du lundi 16 juillet 1827. Also published in the *Bulletin des sciences naturelles,* v.12, 13–16, 1827 and translated into English, rec.434. Published in full, rec.422. What apparently is a Russian translation and also a condensation, rec.455, was not seen.

428. [Espèces de crocodiles dans le Nil]. *Le Globe*. Paris; 1827 December 13; 6(9): 46.

NOTE. Response by Cuvier to a paper read by Geoffroy, *(Mémoire sur une petite espèce de crocodile vivant dans le Nil. . .*), and further comments by Geoffroy, Cuvier, and Mongez, at a meeting of the Académie des sciences, December 10, 1827.

429. Sur le genre de reptiles batraciens Nommé *Amphiuma,* et sur une nouvelle espèce de ce genre (*Amphiuma tridactylum*). Mémoire lu à l'Académie des Sciences, le 13 novembre 1826. *Paris. Muséum national d'histoire naturelle. Mémoires;* 1827; 14: 1–14. 2 pl.

NOTE. Also published separately; [n.p.; n.d.]; 15p. 2pl. 26 cm. Summarized, rec.416, and condensed rec.425.

430. Du canard pie, à pieds demi palmés, de la Nouvelle-Hollande. (*Anas melanoleuca,* Lath.). *Paris. Muséum national d'histoire naturelle. Mémoires;* 1827; 14: 345–347. 1 pl.

NOTE. Published also separately; [n.p., n.d.]; 3p., 1pl., 4°.

431. Rapport . . . sur le [Notice géognostique sur les Terrains secondaires de l'étang de Berre (Bouches-du-Rhône) par MM. Delcros et Rozet], fait à l'Académie Royale des sciences dans la séance du 2 avril

1827. *Société linnéene de Normandie, Caen. Mémoires;* 1827; 1826–1827: 91–96.

NOTE. Report signed by Cuvier.

432. Opinion exprimée, . . . sur le Mémoire précédent de M. Constant Prévost (Extrait de l'analyse des travaux de l'Académie des Sciences, pour l'année 1827, rec.641). In: Prévost, Constant. Documents pour l'histoire des terrains tertiaires. n.p.; [1809–1827]: 125–126.

NOTE. Extract from p. cxxiv–cxxv, cxxvj of *Analyse des travaux. . .* for 1827, rec.641. The Mémoire of Constantin Prévost referred to in the title is *Les continents actuels ont-ils été, à plusieurs reprises, submergés par la mer? Dissertation géologique* . . . , p.123–124.

433. Rapport fait à l'Académie des Sciences en mars 1821, par M. le baron Cuvier . . . In: Serres, Étienne Renaud Augustin. Anatomie comparée du cerveau, dans les quatre classes vertébrés, appliquée à la physiologie et la pathologie du système nerveux; . . . Tome I. Paris and Montpellier: Gabon; Bruxelles: Dépot Général de Librairie Médicale Française; 1827; 1: xviij–xxxviij.

NOTE. Extracts from rec.623 and rec.625. Also published with another report by Cuvier on a work by Serres, rec.361. Translated into German with some condensation, rec.388.

434. Observations on Caverns containing Bones, with an account of the Grotto of Oiselles, near Besançon. *Edinburgh journal of science;* 1827; July/Oct. 7: 279–282.

NOTE. Translation of rec.427, which is a condensation of rec.422.

435. Observations on the Scarus of the Ancients. By Baron Cuvier. *Edinburgh journal of science;* July/Oct. 1827; 7: 355–356.

NOTE. Translation of rec.426.

436. Rapport . . . sur le Mémoire présenté à l'Académie par MM. Audouin et Milne Edwards, et qui a pour objet d'exposer leurs recherches récentes sur les animaux sans vertèbres qu'ils ont observés aux

îles Chausey. *Académie des sciences. Procès verbaux;* 1921; 9(1828–1831): 138–139.

NOTE. Presented at the Nov. 3, 1828 meeting of the Académie and signed by Duméril, and Cuvier as reporter. An earlier and very brief summary appeared in *Le Globe,* rec.449. Condensed and slightly different versions were also published, rec.440 and rec.461, the latter with some material not included in the above.

437. Rapport . . . sur le Mémoire de M. Flourens relatif aux Effets de la section des canaux semi-circulaires dans les jeunes lapins. *Académie des sciences. Procès verbaux;* 1921; 9(1828–1831): 148–149.

NOTE. Presented at the Nov. 24, 1828 meeting of the Académie and signed by Duméril, and Cuvier as reporter. A summary of the same paper appeared in *Le Globe,* rec.508. A report on an earlier memoir on the same general subject was also published, rec.438.

438. Rapport . . . sur un Mémoire de M. Flourens, concernant les effets de la résection des canaux semi-circulaires de l'oreille. *Annales de chimie et de physique;* 1828; 39: 104–109.

NOTE. Report presented by Cuvier at the meeting of Sept. 15, 1828 of the Académie des sciences. Signed by Portal, Duméril, and Cuvier. No reporter is indicated. Also published in the *Académie des sciences, Procès verbaux;* 1921; 9(1828–1831): 117–119; and republished in 1840 in the *Annales des sciences naturelles,* Ser. II, v.13 (Zool): 305–309. A shortened and probably earlier version appeared in *Le Globe,* rec.445. A second memoir on the same general subject was also published, rec.437.

439. Rapport fait à l'Académie des Sciences le 13 octobre 1828, . . . sur le Mémoire de MM. Marcel de Serres, Dubreuil et Jean-Jean, sur les Cavernes de Lunel-Vieil et sur les ossemens d'hyènes qu'elles renferment. *Annales de chimie et de physique;* 1828; 39: 217–222.

NOTE. Signed by A. Brongniart, and Cuvier as reporter. Also published in the *Académie des sciences, Procès verbaux;* 1921;

9(1828–1831): 131–133. Published earlier with some omissions, rec.447, which was translated into Italian, rec.452.

440. Extrait du rapport fait à l'Académie des sciences sur le Mémoire présenté par MM. Audouin et Milne Edwards, dans la séance du 29 septembre 1828, et lu dans celle du 8 octobre; Par MM. Cuvier et Duméril. *Annales des sciences naturelles;* 1828; 15: 111–112.

NOTE. Presented at the November 3, 1828 meeting of the Académie des sciences. A much briefer report than rec.461 which, however, gives above as its source. Published also separately, n.p., n.d., 3p. 22cm. An earlier, but still briefer report was published in *Le Globe,* rec.449. Published more fully, rec.436. Cuvier is listed as reporter in rec.436.

441. Rapport verbal fait à l'Académie des Sciences sur un ouvrage de MM. l'abbé Croiset et Jobert aîné, intitulé: Recherches sur les Ossemens fossiles du département du Puy-de-Dôme . . . (Académie royale des Sciences, séance du 27 octobre 1828). *Annales des sciences naturelles;* 1828; 15: 218–224.

NOTE. Also published separately, [Paris, Imprimerie de C. Thuau, 1828], 8p. 22cm. A continuation of rec.412. Published with omissions in *Le Globe,* rec.448, which was translated into German, rec.454.

442. Rapport fait à l'Académie des sciences sur les Manuscrits et les dessins envoyés du Port-Jackson par MM. Quoy et Gaymard, pour être déposés dans les archives de l'Académie, ainsi que sur les Collections adressées par les mêmes naturalistes au Muséum d'histoire naturelle. *Annales maritimes et coloniales;* 1828; Pt. 2, v.[35]: 387–393.

NOTE. Report presented at the Feb. 11, 1828 meeting of the Académie des sciences. Signed by Duméril, and Cuvier as reporter. Also published in the *Académie des sciences, Procès verbaux;* 1921; 9(1828–1831): 18–20 and reported in *Le Globe,* rec.443.

443. [. . . Rapport sur les manuscrits et les dessins envoyés du Port Jackson par MM. Quoy et Gaymart pour être déposés dans les archives

de l'Académie . . .]. *Le Globe*. Paris; 1828 February 16; 6(33): 204–205.

NOTE. Report made by Duméril and Cuvier, and read by Cuvier; presented at the Académie des Sciences at the February 11, 1828 meeting. A more extensive report of the same paper was also published, rec.442.

444. [Comments on a memoir of M. Chevreul] . . . sur l'influence que deux couleurs peuvent avoir l'une sur l'autre quand on les voit simultanément, et sur la nécessité de prendre cette influence en considération pour juger des couleurs, abstraction faite sur leur solidité. *Le Globe*. Paris; 1828 April 16; 6(60): 343.

NOTE. Comments made at the meeting of the Académie des sciences on April 7, 1828 following M. Chevreul's presentation.

445. Rapport sur les dernières expériences de M. Flourens. *Le Globe*. Paris; 1828 September 27; 6(97): 720.

NOTE. Presented September 15, 1828 at a meeting of the Académie des sciences. Signed by MM. Portal, Duméril, and Cuvier, who was reporter. Rec.438 is a more complete version.

446. [Report by Cuvier and Duméril at the October 6, 1828 meeting of the Académie des sciences on the *Résumé de leurs recherches sur les animaux sans vertèbres faites aux îles Chausey* by Audouin and Milne-Edwards]. *Le Globe*. Paris; 1828 October 15; 6(102): 761–762.

NOTE. Very brief report on a résumé by Audouin and Milne Edwards of a paper they later gave in full, which was also reported on by Cuvier and Duméril, rec.436, rec.440, and rec.461. Cuvier is given as reporter in rec.436.

447. Sur les cavernes de Lunel-Vieil, et les ossements fossiles qui s'y trouvent. *Le Globe*. Paris; 1828 October 18; 6(103): 767.

NOTE. Report on paper by Marcel de Serres made by Cuvier and Brongniart to the Académie des sciences at the meeting of October 13, 1828 with Cuvier as reporter. Translated into Italian, rec.452. A more complete version was also published, rec.439.

448. Rapport verbal sur le premier volume de l'ouvrage intitulé: *Recherches sur les ossements fossiles du département du Puy de Dôme,* par MM. l'abbé Croiset et Jobert aîné. *Le Globe.* Paris; 1828 November 1; 6(107): 799.

NOTE. Read at the October 27, 1828 session of the Académie des sciences. Published more fully, rec.441. Translated into German, rec.454.

449. [Rapport sur un mémoire de MM. Audouin et Milne-Edwards, qui contient le résumé des observations faites sur les animaux marins des côtes de la Manche]. *Le Globe.* Paris; 1828 November 5; 6(111): 831–832.

NOTE. Presented to the Académie des sciences at the meeting of November 5 [i.e.3?], 1828 by Cuvier in his name and that of Duméril. Above is a still briefer summary than rec.440 or rec.461 of the report as it was published in 1921, rec.436. Cuvier is listed as reporter in rec.436.

450. [Speech on annual revision of electoral laws]. *Le Moniteur universel.* Paris; 1828 May 2; Suppl. to N°123: 548.

NOTE. Chambre des Députés, Session of April 1828.

451. Rapport verbal fait à l'Académie royale des Sciences, par M. Cuvier, sur l'histoire naturelle des oiseaux de l'Amérique septentrionale de M. Audubon. *Le Moniteur universel.* Paris; 1828 October 1; (275): 1528.

NOTE. Read at the meeting of the Académie des sciences on September 22, 1828.

452. Su lè Caverne di Lunel-Vieil, e su gli ossami fossile che vi si trovano. In: Bertrand, Alexandre Jacques F. Lettere su le rivoluzione del globo . . . Parigi; 1828: 138–141.

NOTE. Bound with Cuvier, G. Discorso . . . su le rivoluzioni . . . v.2, rec.688. Translation of rec.447, a report by A. Brongniart and Cuvier as reporter, on above paper by Marcel de Serres. The report was presented at the meeting of October 13, 1828 to the Académie des sciences.

453. Ueber die Naturgeschichte der Fische. Von Cuvier und Valenciennes. [Prospectus by Cuvier]. *Notizen aus dem Gebiete der Natur= und Heilkunde,* Erfurt; 1828; 20: col. 241–250.

NOTE. Translation of rec.424 with omission of section on authors' indebtedness to individual travelers.

454. Ueber die fossilen Knochen im Berge Périer hat Baron Cuvier den 27. October 1828 der Akademie der Wissenschaften Bericht erstattet. *Notizen aus dem Gebiete der Natur= und Heilkunde,* Erfurt; 1828; 22: col. 277–279.

NOTE. Translation of rec.448 with omission of introductory paragraph.

455. O Kostiakh prezhdepotopnykh zhivotnykh, vnov' otkrytykh bliz Bezansona. *Vestnik estestvennykh nauk i meditsiny;* 1828; 2(5): 12–15.

NOTE. Not seen and not located outside of the former USSR. The Akademiia Nauk, spravochno bibliograficheskii otdel, Biblioteka, St. Petersburg, has a copy. Possibly a translation of rec.422 or rec.427.

456. Rapport . . . sur le troisième *Mémoire zoologique* de MM. Quoy et Gaymard. *Académie des sciences. Procès verbaux;* 1921; 9(1828–1831): 188–190.

NOTE. Presented at the Feb. 2, 1829 meeting of the Académie and signed by Duméril, and Cuvier as reporter. A report by Cuvier and Duméril on the fourth and fifth memoirs was also published, rec.460.

457. Rapport verbal sur la monographie des Aplysiens de M. Rang, fait à l'Académie royale des Sciences. *Annales des sciences naturelles;* 1829; 16: 300–303.

NOTE. Read at the February 23, 1829 meeting of the Académie des sciences. Published also in 1831 in the *Bulletin des sciences naturelles*, v.24: 375–377.

458. Rapport fait à l'Académie des Sciences sur la partie zoologique du Voyage de la Chevrette dans les mers de l'Inde . . . (Séance du 6 [i.e.9] mars 1829). *Annales des sciences naturelles;* 1829; 16: 331–336.

NOTE. Report signed by Desfontaines, Geoffroy Saint Hilaire, Duméril, and Cuvier as reporter. Also published in the *Bulletin des sciences naturelles,* 1829 17: 199–203 and again in v.19: 330–334, 1829, of the same journal; in the *Annales maritimes et coloniales* 1829 v.[38], Pt.2, T.1: 309–314; in the *Académie des sciences, Procès verbaux;* 1921; 9(1828–1831): 203–205. It was also published separately, [Paris?, Imprimerie de C. Thuau, n.d.], 7p. 21cm. A German summary was also published, rec.555.

459. Mémoire sur un ver parasite d'un nouveau genre (*Hectocotylus octopodis*); . . . (Lu à l'Académie royale des Sciences, le 12 octobre 1829). *Annales des sciences naturelles;* 1829; 18: 147–156, 1 pl.

NOTE. Also published separately, [Paris, Imprimerie de C. Thuau, n.d.] 10p. 1 pl. Translated into German, rec.522. Extracted and condensed, rec.462. A shortened version appeared in *Le Globe,* rec.466, and was translated into English, rec.475. A condensed German translation, rec.552, is not a translation of the *Globe* article nor of rec.462.

460. [Rapport fait à l'Académie des sciences, . . . sur le quatrième et le cinquième mémoire de zoologie envoyés par MM. Quoy et Gaymard, naturalistes de l'Astrolobe]. *Annales maritimes et coloniales;* 1829; Pt.2(2): [39]: 27–30.

NOTE. Presented at the May 25, 1829 meeting of the Académie des sciences and signed byDuméril, and Cuvier as reporter. Published also in the *Académie des sciences, Procès verbaux;* 1921; 9(1828–1831): 253–254; and summarized, rec.473. A report on the third memoir was also published, rec.456.

461. Extrait du rapport fait sur ce Mémoire à l'Acad. roy. des sciences, par MM. Cuvier et Duméril. *Bulletin des sciences naturelles;* 1829; 17: 127–129.

NOTE. Report referred to in title is the "Mémoire présenté par MM. Audouin et Milne Edwards sur leurs recherches sur les animaux sans vertèbres qu'ils ont observés aux îles Chausey." Above is a condensation of Cuvier and Duméril's report, as published in 1921, rec.436, presented at the November 3, 1828 meeting of the Académie des sciences, although it contains some material not in rec.436. In spite of listing rec.440 as its source, above is a different reporting. An earlier condensation also appeared, rec.449. Cuvier is given as reporter in rec.436.

462. *Hecatostoma,* nouveau genre de ver parasite; Mém. lu à l'Académie des Sciences dans la séance du 11 [i.e.12] octobre 1829. *Bulletin des sciences naturelles;* 1829; 19: 409–411.

NOTE. Extracts, condensation, and rearrangement of rec.459. Another report on same paper was also published, rec.466.

463. Rapport . . . sur les travaux de M. Ad. Brongniart sur les végétaux fossiles. *Le Globe;* 1829 January 7; 7(2): 14–15.

NOTE. Presented at the Dec. 29, 1828 meeting of the Académie des sciences and signed by Mirbel, Fourier, Beudant, and Cuvier as reporter. Also published in *Académie des sciences, Procès verbaux;* 1921; 9(1828–1831): 170–172.

464. Rapport sur les collections rapportés d'Égypte par M. Rifaud. *Le Globe*. Paris; 1829 March 29; 7(25): 199–200.

NOTE. Report made at the Meeting of the Académie des sciences, March 16, 1829 by Cuvier in his name and that of Desfontaines, Geoffroy St. Hilaire, Labilliardière, Duméril, and Blainville. Published more fully, rec.472.

465. Rapport de M. G. Cuvier sur le mémoire de M. Roulin relatif à une nouvelle espèce de tapir. *Le Globe*. Paris; 1829 April 18; 7(31): 247.

NOTE. Read April 12, 1829 at the session of the Académie des sciences. A more complete text was published later, rec.471.

466. [Mémoire sur un nouveau genre de ver parasite. *Hecatostoma*]. *Le Globe;* 1829 Oct. 14; 7(82): 653–654.

NOTE. Condensed report of rec.459 almost entirely in Cuvier's words. Read at the Académie des sciences at the meeting on October 11 [i.e.12], 1829. Translated into English, rec.475. Another report on same paper was also published, rec.462.

467. Rapport sur les collections zoologiques recueillies par MM Quoy et Gaimart dans le cours de l'expédition commandée par M. Durville. *Le Globe*. Paris; 1829 Nov. 4; 7(88): 700.

NOTE. Report made at the meeting on October 26, 1829 of the Académie des sciences by M. Cuvier in the name of the natural history section.

468. [Remarques sur une machoire de palæotheriums . . .]. *Le Globe*. Paris; 1829 Nov. 11; 7(90): 713.

NOTE. Remarks made during a meeting of the Académie des sciences, November 9, 1829.

469. Rapport sur le Voyage de M. Charles Belanger . . . [sur les collections recueillies par M. Bélanger dans son voyage aux Indes]. *Le Moniteur universel,* Paris; 1829 October 5; (278): 1617–1618.

NOTE. Presented at the meeting of Sept. 28, 1829. Signed by Geoffroy St. Hilaire, Latreille, Duméril, Desfontaines, Mirbel, H. Cassini, with Cuvier as reporter. Also published in *L'Universel,* 1829 October 7; 1(280): 339–340; in *Bulletin des sciences naturelles,* 1829; 19: 223–320; *Annales maritimes et coloniales,* 1829, Pt.2 (2): [39]: 652–660; and in the *Académie des sciences, Procès verbaux;* 1921; 9(1828–1831): 322–325. There are some differences in the numbers of species reported among the several publications.

470. Discours prononcé . . . , à l'assemblé générale annuelle du 5 décembre 1828 de la Société de géographie. *Société de géographie. Bulletin;* 1829; 11: 29–32.

NOTE. Published also in *Annales maritimes et coloniales;* 1829, Pt.2(1); v.[38]: 342–344.

471. Rapport fait à l'Académie des sciences, le 13 avril 1829, par MM. G. Cuvier et Duméril, sur le travail de M. le docteur Roulin intitulé: "Mémoire pour arriver à l'histoire des Tapirs, et description d'une nouvelle espèce appartenant aux hautes régions de la Cordillière des Andes, avec des considérations sur les animaux fabuleux dont l'histoire se rapporte à celle du Tapir," (lu à l'Académie, le 9 fevrier 1829). *L'Universel.* Paris; 1829 April 30; (120): 411–412.

NOTE. Read at the Académie des sciences April 13, 1829. Signed by Duméril, and Cuvier as reporter. Also published in *Annales des sciences naturelles;* 1829; 17: 107–112; in *Annales de chimie et de physique;* 1829; Ser.II 40: 416–422; in *Bulletin des sciences naturelles;* 1829; 17: 118–123; *Académie des sciences, Procès verbaux;* 1921; 9(1829–1831): 224–226; and separately, [Paris, Imprimerie de Firmin Didot n.d.], 7p. 8°. Extracts were also published, rec.465. Translated into English with some condensation, rec.477; and condensed and translated into German, rec.556.

472. Rapport sur les collections et les dessins d'histoire naturelle rapportés d'Égypte et de Nubie par M. Rifaud. Extrait du procès verbal de la séance du lundi 16 mars 1829. *L'Universel.* Paris; 1829 May 4/5; (124/125): 426–427.

NOTE. Presented at the March 16, 1829 meeting of the Académie des sciences and signed by Desfontaines, de La Billardière, Geoffroy Saint-Hilaire, Duméril, de Blainville, H. de Cassini, and Cuvier as reporter. Published also in *Annales maritimes et coloniales;* 1829; Pt.2 1(38): 564–568; in the *Académie des sciences, Procès verbaux;* 1921; 9(1828–1831): 208–210; and separately, [Paris, Imprimerie de Firmin Didot], n.d., 6p. 21.5cm. Extracts were published March 29, 1829, rec.464.

473. Rapport sur les 4e. et 5e. mémoires de zoologie envoyés à l'Académie par MM. Quoy et Gaymard. *L'Universel.* Paris; 1829 May 27; (147): 501–502.

NOTE. Presented at the May 25 meeting of the Académie des sciences and signed by Duméril, and Cuvier as reporter. Above is a summary of the report which was later published more fully, rec.460.

474. Discours de M. le baron Cuvier, Directeur, Sur les prix de vertu. [Paris]: Imprimerie de A. Firmin Didot; [1829?]. 8p. 4°.

NOTE. At head of title: Institut Royal de France. Académie française. Also published on p.[9]–27 of *Prix de vertu, Fondé par M. de Montyon. Discours prononcé par M. le Baron Cuvier, dans la séance publique de la Saint-Louis, 1829, sur les prix de vertu décernés dans cette séance . . . ,* Paris, A. Firmin Didot, 1829; and again in *Recueil des discours, rapports et pièces diverses lus dans les séances publiques et particulières de l'Académie Française, 1820–1829,* Paris, Firmin Didot, 1843: 667–677, which gives the date of August 25, 1829 for the discourse. Published in 1858 in Lock, Frédéric and J. Couly d'Aragon, *Les Prix de vertus fondés par M. de Montyon, . . . ;* Paris, Garnier frères, 1858; v.1, 137–147.

475. *Hecatostoma* (*Hecatoncotylus*), a new genus of parasitical worms, described by M. Cuvier, before the Academy of Paris, 11th [i.e.12th] October, 1829. *Edinburgh journal of natural and geographical science;* Oct. 1829/Mar.1830; 1(Nov): 218–220.

NOTE. Translation of rec.466. Published also in the *Edinburgh philosophical journal;* 1830; 8: 101–104.

476. On the state of Natural History, and the progress which it has made since the return of the Maritime Peace. *Edinburgh new philosophical journal;* April/Oct., 1829; [7]: 1–14.

NOTE. Translation of rec.402 with some omissions.

477. On a new Species of Tapir, discovered in the Andes, with Remarks on Antediluvian Animals, [Report on]. *Edinburgh new philosophical journal;* April/Oct. 1829; [7]: 155–159.

NOTE. Translation, with some condensation, of a report, rec.471, on a paper by Roulin. The report was presented to the Académie des sciences on April 13, 1829 with Cuvier as reporter.

478. On the right Use of Generic Names in Natural History; according to the Opinions of MM. Cuvier and DeCandolle. To the Editors of the Philosophical Magazine, and Annals. *Philosophical magazine,* London; 1829 July/Dec.; n.s.6: 348–351.

NOTE. Cuvier's letter appears on p.348, followed by "Extract from the Preface to the First Edition of Cuvier's *Règne animal*" (rec.747) in French, with English translation, p.349–351.

479. Allgemeine Charactere und wesentliche Natur der Fische. Von Cuvier. (Ein Capitel aus der *Histoire naturelle des Poissons* Par Mr. le Bar. Cuvier et Mr. Valenciennes. Paris, 1828 . . .). *Notizen aus dem Gebiete der Natur= und Heilkunde,* Erfurt; 1829; 23: col.164–169.

NOTE. Translation of *Histoire naturelle des poissons,* v.1, 1828, p.271–285, 8° printing, rec.706, and p.199–201 of the 4° printing, rec.707.

480. Cours d'histoire des sciences naturelles. *Le Globe*. Paris; 1829 Dec. 30; 7: 825–827; 1830 Jan. 6: 12–14; 1830 Jan. 13: 26–28; 1830 Jan. 20: 44–46; 1830 Jan. 27: 58–60; 1830 Feb. 3: 79–80; 1830 Feb. 10: 92–94; 1830 Feb. 19: 17–18; 1830 Feb. 21: 25–27; 1830 Feb. 27: 49–52; 1830 March 3: 73–76; 1830 March 13: 105–108; 1830 March 22: 141–144; 1830 April 19: 249–252; 1830 April 26: 277–279; 1830 April 28: 285–287; 1830 May 4: 309–311; 1830 May 7: 321–322; 1830 May 20: 373–376; 1830 May 23: 384–388; 1830 May 29: 409–411; 1830 June 6: 438–439; 1830 June 23: 507–509; 1830 July 25: 633–635.

NOTE. Also published in a more condensed reporting in *Le Temps,* rec.483, which continues to final lecture in 1832. Occasionally somewhat different information is reported. Other reports were also published, rec.481, rec.482, rec.483, rec.484. Above is a different reporting from that of *Histoire des sciences naturelles . . . rédigé . . . par M. Magdeleine de Saint-Agy* (rec.702), but only some lectures were compared. Some of the lectures were translated into English, rec.485 and rec.486, and into German, rec.487. See also the Note to rec.483 for more information on the publication of these lectures.

481. Cours de M. Cuvier au Collège de France. Histoire des sciences naturelles. *Le National.* Paris; 1830 Jan. 1: 1–2; 1830 Jan. 7: 1–3; 1830 Jan. 8: 2–3; 1830 Jan. 17: 1–3; 1830 Jan. 27: 1–3; 1830 Feb. 6: 2–4; 1830 Feb. 11: 1–3; 1830 Feb. 26: 1–2; 1830 March 8: 1–3; 1830 March 31: 1–2; 1830 April 15: 2–4; 1830 May 6: 1–3.

NOTE. A condensed reporting of the lectures. For further information on the publication of these course lectures, see Notes to rec.480 and rec.483.

482. Cours d'histoire des sciences naturelles. *Revue encyclopédique;* 1830 July/Sept.; 47: 5–24.

NOTE. Condensations and extracts by Godinet of Cuvier's lectures. For further information on the publication of these lectures, see Notes to rec.480 and rec.483.

483. Cours de M. Cuvier. Histoire des sciences naturelles. *Le Temps.* Paris; 1829 Dec. 16: cols.733–740; 1829 Dec. 20: cols.783–789; 1829 Dec. 24: cols.829–838; 1829 Dec. 27: cols.856–860; 1829 Dec. 28: cols.865–871; 1829 Dec. 30: cols.892–899; 1830 Jan. 3: cols.943–948; 1830 Jan. 6: cols.976–981; 1830 Jan. 8: cols.997–1002; 1830 Jan. 10: cols.1021–1026; 1820 Jan. 11: cols.1033–1039; 1830 Jan. 13: cols.1057–1062; 1830 Jan. 16: cols.1093–1098; 1830 Jan. 18: cols.1117–1121; 1830 Jan. 20: cols.1141–1147; 1830 Jan. 21: cols.1160–1166; 1830 Jan. 22/23: cols.1172–1178; 1830 Jan. 24: cols.1184–1188; 1830 Jan. 29: cols.1243–1248; 1830 Feb. 1: cols.1275–1280; 1830 Feb. 4: cols.1311–1316; 1830 Feb. 7: cols.1347–1351; 1830 Feb. 10: cols.1383–1389; 1830 Feb. 17: cols.1472–1478; 1830 Feb. 20: cols.1509–1514; 1830 Feb. 21: cols.1519–1525; 1830 March 1: cols. 1616–1621; 1830 March 3: cols.1635–1640; 1830 March 10: cols.1719–1725; 1830 March 15: cols.1781–1784; 1830 March 21: cols.1851–1856; 1830 March 26: cols.1918–1921; 1830 March 27: cols.1923–1925; 1830 April 2: cols.1995–1997, 2001–2003; 1830 April 6: cols.2049–2054; 1830

April 14: cols.2127–2134; 1830 April 18: cols.2183–2186; 1830 April 20: cols.2202–2207; 1830 April 22: cols.2230–2234; 1830 April 23: cols.2239–2245; 1830 April 25: cols.2264–2269; 1830 April 26: cols. 2271–2276; 1830 April 28: cols.2298–2305; 1830 April 29: cols.2312–2316; 1830 May 1: cols.2331–2334; 1830 May 2: cols.2348–2352; 1830 May 5: cols.2384–2390; 1830 May 9: cols.2432–2438; 1830 May 13: cols.2479–2485; 1830 May 19: cols.2554–2558; 1830 May 23: cols. 2597–2604; 1830 May 27: cols.2648–2544[i.e.2654]; 1830 May 30: cols.2682–2688; 1830 June 2: cols.2708–2714; 1830 June 6: cols.2753–2759; 1830 June 12: cols.2824–2830; 1830 June 19: cols.2912–2917; 1830 June 21: cols.2935–2942; 1830 June 24: cols.2969–2975; 1830 June 27: cols.3007–3013; 1830 July 1: cols.3054–3060; 1830 July 9: cols.3147–3153; 1830 July 17: cols.3246–3253; 1831 Feb. 13; (485): cols.6617–6619; 1831 April 25; (555): cols.8072–8073; 1831 May 1; (559): cols.8191–8193; 1831 May 13; (571): cols.8413–8415; 1831 June 14; (604): cols.8964–8970; 1831 June 24; (614): cols.9095–9103; 1831 July 20; (640): cols.9489–9497; 1832 June 5; (961): cols.14,445–14, 454.

NOTE. Includes Cuvier's last lecture, which was not found in any other publication consulted. It was apparently also published separately under the title *Histoire des sciences naturelles . . . Dernière leçon prononcée le 8 mai 1832*, Paris, Bechet, 1832, but this has not been located. The latter is cited in the *Bibliographie de la France*, v.21 (22), 1832, #2637. Many issues of *Le Temps* are missing from the microfilm examined, especially after June 1831. The overall title under which the course is published varies. The above is a different reporting from that in *Le Globe*, rec.480, which is not carried beyond 1830. *Le Temps*, above, reporting is for the most part a more condensed version, although occasionally it has different information. These lectures were also published in 5 volumes, rec.702. The above has slightly different wording and some condensation. Only a few lectures were compared. See rec.702 for other monographic publications and translations. The lectures were published with varying degrees of completeness. See Note to rec.480 for further information.

484. Cours sur l'histoire des sciences naturelles. Par M. le Baron Cuvier (Au Collège de France). *L'Universel.* Paris; 1829 Dec. 17: 67; 1830 Jan. 18: 191–192; 1830 Jan. 30: 235–236; 1830 Feb. 6: 263–264; 1830 Feb. 26: 343–344; 1830 March 9: 387–388; 1830 March 22: 438–439; 1830 April 24: 567–568; 1830 May 11: 44.

NOTE. Not compared with other publications of this series of lectures. A German translation, rec.487, cites the lecture reported above for Jan. 18, 1830 as its source. For further information on the publication of these lectures, see Notes to rec.480 and rec.483.

485. Course of Lectures on the History of the Natural Sciences; by Baron Cuvier. *Edinburgh journal of natural and geographical science;* Oct.1829/Mar.1830; 1: 457–461; Apr.1830/Sept.1831; 2: 51–55, 132–140, 213–222, 285–292, 369–372, 446–451; n.s.1 [i.e.3]: 20–26, 89–96.

NOTE. "... [we] present to our readers a series of abstracts of these lectures, . . . which will be condensed in each successive Number . . ., —Ed.;" v.1, p.457. "... the reports ... being drawn from *Le Globe* newspaper, must be viewed in the light of mere epitomes of the eloquent details which compose the course of M. Cuvier; but from a comparison with still more abstracted reports contained in the *Revue encyclopédique* [rec.482], we feel authorized in considering that we give a sufficiently extensive view of the subject. Ed."; v.3, p.20. A comparison with *Le Globe,* rec.480, up to the 10th article shows above to be a direct translation. The above is not the same English translation as that appearing in the *Edinburgh new philosophical journal,* rec.486. The above is not a translation of the *Revue encyclopédique,* rec.482, which includes some material not in above and which omits other parts. For further information on publication of these lectures, see Notes to rec.480, rec.483, and rec.702.

486. Lectures on the History of the Natural Sciences. By Baron Cuvier. *Edinburgh new philosophical journal;* Oct.,1829/March, 1830; 8: 326–349; 9: 41–83; 10: 49–58.

NOTE. "Of these highly interesting lectures, at present delivering in Paris, we shall in this and succeeding no's. give such as will interest our readers. The reports we now publish are held by some of our friends now attending these lectures to be correct." Footnote, p.326. Translation of some of the lectures which appeared in *Le Globe,* rec.480. The above is not the same translation as rec.485. For further information on the publication of these lectures, see Notes to rec.480 and rec.483.

487. Vorlesungen über die Geschichte der Naturwissenschaften Von dem Baron Cuvier. *Notizen aus dem Gebiete der Natur= und Heilkunde,* Erfurt; 1829 [published in 1830?]; 26(10): col.145–150; 292–295.

NOTE. Translated excerpts from Cuvier's lectures on natural history. Above cites as its source *L'Universel,* 18 January 1830, rec.484. Seen but not compared. For further information on the publication of these lectures, see the Notes to rec.480 and rec.483.

Records 488–502 for the Geoffroy Saint Hilaire debate are arranged by date of presentation to the Académie.

488. Considérations sur les Mollusques, et en particulier sur les Céphalopodes. . . . (Lues à l'Académie royale des Sciences, février 1830). *Annales des sciences naturelles;* 1830; 19: 241–259.

NOTE. Read at the session of February 22, 1830 at the Académie. Published also separately, with a few minor but interesting alterations [Paris: Crochard, n.d.]; 18p. 1pl. 8°. Published with some of these minor alterations in the *Revue encyclopédique,* 46: 5–20, April/June 1830. Different condensed versions of the memoir were also published in newspapers, rec.490, rec.491, and rec.492, and therefore, probably earlier. Another condensed version appeared in a journal, rec.489. Translated into English with condensations and omissions, rec.493.

489. Considérations sur l'unité de composition en général et sur la structure des Céphalopodes en particulier, lues à l'Académie des sci-

ences, le 22 février 1830. *Bulletin des sciences naturelles;* 1830; 20: 304–311.

NOTE. Condensation plus extracts of Cuvier's paper, which was published more fully in rec.488. Except for introductory section, text is same as that published in *Le Globe,* rec.490.

490. Considérations sur les mollusques, et en particulier sur les céphalopodes. *Le Globe*. Paris; 1830 Feb. 24; 6e. année(10): 37–40.

NOTE. Condensation and extracts from a memoir read by Cuvier at the February 22 meeting of the Académie des sciences, which was more fully reported in rec.488. First published incompletely in rec.491. Except for introductory section, text of above is same as that of rec.489. An English translation, rec.493, cites this as source, but the text is somewhat different.

491. [Considérations sur les mollusques, et en particulier sur les céphalopodes]. *Journal des débats*. Paris; 1830 Feb. 23: [2–3].

NOTE. Read at the Académie des sciences at the session of February 22, 1830. Also published in E. Geoffroy Saint-Hilaire, *Principes de philosophie zoologique discutés en mars 1830 . . . ,* Paris, Pichon et Didier, 1830, p.53–72, with introduction and extensive notes by Geoffroy. The latter was reproduced in Patrick Tort, *La Querelle des analogues . . .* Plan de la Tour (Var.), Editions d'aujourd'hui, [1983], p.123–142. Published more fully in rec.488.

492. . . . Considérations sur les Mollusques, et en particulier sur les céphalopodes. *L'Universel*. Paris; 1830 Feb. 24 and 25; (55): 336 and (56): 339–340.

NOTE. Not compared with other publications of this memoir. Read February 22 at the session of the Académie des sciences.

493. Considerations on the Mollusca, and on the Cephalopoda in particular. *Philosophical magazine,* London; 1830 Jan./June; 7: 381–386.

NOTE. Translation which is a condensation with omissions and extracts of rec.488. Although *Le Globe,* rec.490, is acknowledged as source for this, the translation does not follow it exactly.

494. Considérations sur l'os hyoïde. *Le Globe*. Paris; 1830 March 23; (37): 145–148.

NOTE. Read at the Académie des sciences on March 22, 1830. Rec.495 is a different reporting of the same paper with different omissions and inclusions. Continued by rec.500.

495. Considérations sur l'os hyoïde. *Journal des débats*. Paris; 1830 March 23: [3].

NOTE. Read at the session of the Académie des sciences on March 22, 1830. Another reporting of rec.494. Continued by rec.498. Published also in the *Revue bibliographique;* 1830 [v.1]: 33–39 (with: *Annales des sciences naturelles,* 1830, v.19). Published again in E. Geoffroy Saint-Hilaire, *Principes de philosophie zoologique discutés en mars 1830 . . .* , Paris, Pichon et Didier, 1830, p.139–162, with introduction and extensive footnotes by Geoffroy. The latter work was reproduced in Patrick Tort, *La Querelle des analogues . . .* Plan de la Tour (Var.), Editions d'aujourd'hui, [1983] p.209–232. Another reporting was also published, rec.494. Extracts and condensations of the above were also published, rec.496 and rec.501, the latter with introductory commentary and E. Geoffroy Saint-Hilaire's reply.

496. [Mémoire sur l'os hyoïde]. *Le National*. Paris; 1830 March 23: 2–4.

NOTE. Read at the session of March 22, 1830 at the Académie des sciences. An extract of rec.495. Published also in *Gazette médicale de Paris;* 1830 March 27 1(13): 118–119.

497. [Mémoire sur le sternum]. *Gazette médicale de Paris;* 1830 April 10; 1(15): 135–136.

NOTE. Read at the April 5, 1830 session of the Académie des sciences. Published in *Le National* on April 6, rec.500, with additional material. Above is not the same reporting as either rec.498 or rec.499.

498. Considérations sur le sternum. *Le Globe*. Paris; 1830 April 6; 6(51): 201–202.

NOTE. Discussion and review of the "polémique" between Cuvier and Geoffroy with quotations from Cuvier's paper presented to the Académie des sciences on April 5, 1830, and reported more fully in rec.497 and rec.499. Another reporting was also published, rec.500.

499. Suite des considérations sur l'os hyoïde . . . Digression sur le sternum. *Journal des débats*. Paris; 1830 April 6: [1–3].

NOTE. Continuation of rec.495. Read at the April 5, 1830 session of the Académie des sciences. Also published in the *Revue bibliographique;* 1930; v.[1]: 43–47, (with: *Annales des sciences naturelles,* v.19, 1830). Published with some differences in *Le Globe,* rec.498. Not same reporting as rec.497 and rec.500.

500. [Le sternum chez les quadrupèdes, dans les ovipares, dans les tortues, . . .]. *Le National*. Paris; 1830 April 6; (94): 2–4.

NOTE. Read at the April 5, 1830 meeting of the Académie des sciences. Continues the memoir on the hyoid bone, rec.494 and rec.495. Also published with some omission in rec.497 and with somewhat different reporting, rec.498 and rec.499.

501. [Mémoire sur l'os hyoïde]. *Le Temps*. Paris; 1830 April 7: cols. 2055–2061.

NOTE. A condensation and different reporting of same paper published in rec.494 and rec.495; with introductory commentary, and followed by E. Geoffroy Saint Hilaire's reply.

502. . . . Considérations sur le sternum. *L'Universel*. Paris; 1830 April 7 and 8; (97): 504 and (98):508.

NOTE. Read at the meeting of the Académie des sciences on April 5, 1830. Not compared with other publications of this paper.

503. Rapport fait à l'Académie royale des sciences, séance du 6 septembre 1830 sur un mémoire de M. Breschet, sur les organes de l'audition des poissons. *Annales des sciences naturelles;* 1830; 21: 105–112.

NOTE. Signed by Duméril, and Cuvier as reporter. Published also in *Académie des sciences, Procès verbaux;* 1921; 9(1828–1831): 497–500.

504. Rapport sur trois Mémoires de MM. Victor Audouin et Milne-Edwards, relatifs aux animaux sans vertèbres des côtes de la France; (Fait à l'Académie royale des Sciences, séance du 15 novembre 1830.). *Annales des sciences naturelles;* 1830; 21: 317–329.

NOTE. Signed by Latreille, Duméril, and by Cuvier as reporter. Published also in the *Bulletin des sciences naturelles;* 1831; 24: 89–97; the *Académie des sciences, Procès verbaux;* 1921; 9(1828–1831): 532–536; and separately [Paris?]; Imprimerie de Ve. Thuau; [1830?], 13p., 21cm. Earlier and brief reports were also published, rec.512 and rec.515. Extracts were translated into English, rec.519.

505. Rapport fait à l'Académie royale des Sciences, sur les Collections rapportées récemment de la mer des Indes par M. Dussumier, de Bordeaux; . . . (Lu dans la séance du 13 décembre 1830.). *Annales des sciences naturelles;* 1830; 21: 458–468.

NOTE. Published also in the *Annales maritimes et coloniales;* 1831; Pt.2(1), 45: 281–289, with minor variations. A greatly condensed version appeared January 1, 1831, rec.531.

506. Note sur quelques ossemens qui paraissent appartenir au Dronte, espèce d'oiseau perdue seulement depuis deux siècles; lue à l'Académie des sciences, séance du 12 juillet 1830. *Bulletin des sciences naturelles;* 1830; 22: 122–125.

NOTE. Includes Blainville's response to paper and Cuvier's response to Blainville, as well as remarks by E. Geoffroy Saint Hilaire. Also published on July 18, 1830, but not compared, rec.517. Translated into English, rec.518. Different versions of same report with some omission also appeared, rec.507 and rec.513. A German condensation was also published, rec.535.

507. Note sur quelques ossemens qui paraissent appartenir à une espèce perdue seulement depuis deux siècles. *Le Globe*. Paris; 1830 July 15; 6: 596–597.

NOTE. Read at the Académie des sciences on July 12, 1830. Includes E. Geoffroy's reply. Also published in *Revue bibliothèque;* 1830; [1]: 103–104 (with: *Annales des sciences naturelles;* 1830; 21). Different reports of the same paper were also published, rec.506, rec.513, and rec.517, as well as an English translation, rec.518, and a condensation in German, rec.535.

508. Rapport sur le second mémoire de M. Flourens relatif aux canaux semi-circulaires de l'oreille. *Le Globe*. Paris; 1830 December 3; 6(116): 871.

NOTE. Reported by Cuvier at the session of the Académie des sciences on November 24, 1828. A very brief summary of rec.437.

509. [Discours par M. le baron Cuvier, au nom du conseil royal de l'instruction publique]. *Le Moniteur universel*. Paris; 1830 Jan. 1; (1): 2.

NOTE. Discourse addressed to the King, Charles X.

510. [Rapport fait à l'Académie royale des Sciences sur les travaux des naturalistes attachés à l'expédition du Capitaine d'Urville, dans sa séance du lundi 26 octobre 1829]. *Le Moniteur universel,* Paris; 1830 January 15; (15): 59–60.

NOTE. Report signed by Geoffroy St. Hilaire, Latreille, Duméril, and Cuvier as reporter. Published separately, Paris[?], A. Pihan Delaforest, [n.d.] 8p. 8°; and in *Annales des sciences naturelles,* 1830, 19: 287–297; in *Annales maritimes et coloniales,* 1830, Pt.2(1), [42]: 104–112; in *Académie des sciences, Procès verbaux;* 1921; 9(1828–1831): 343–346. A brief report by Cuvier on the collections was also published, rec.467.

511. Rapport . . . sur le mémoire de M. Bennati relatif *au mécanisme de la voix humaine dans le chant.* *Le National*. Paris; 1830 May 12; (130): 1–2.

NOTE. Report read by Cuvier, as reporter, in his name and those of Prony and Savart at the meeting of May 10, 1830 of the Académie des sciences. Also published in *Gazette médicale de Paris;* 1830 May 15; 1(20): 182. Published more fully later, rec.549. Different condensations appeared in other newspapers: *Le Globe;* 1830 May 12: 341 and *L'Universel;* 1830 May 19(139): 76.

512. [Rapport sur trois mémoires de MM. Audouin et Milne Edwards, relatifs aux animaux sans vertèbres des côtes de la Manche]. *Le National*. Paris; 1830 November 21; (323): 1.

NOTE. Report by Cuvier at the November 15 session of the Académie des sciences. A much fuller report was published later, rec.504, and another condensation was also published in *Le Temps,* rec.515.

513. [Mémoire sur quelques ossemens appartenant au dronte]. *Le Temps*. Paris; 1830 July 14: cols.3208–3209.

NOTE. Condensation of paper read by Cuvier at the July 12, 1830 meeting of the Académie des sciences. An even more condensed version also appeared in *Le National,* July 16, 1830 (no.195), p.[2]. It was reported more fully in the *Le Globe,* rec.507, and in rec.506.

514. [Sur le *dronte* ou *dodo*]. *Le Temps*. Paris; 1830 September 11: cols.3957–3958.

NOTE. Cuvier's response to a memoir read by Blainville at the August 30 meeting of the Académie des sciences.

515. Rapport . . . sur trois mémoires de MM. Audouin et Milne Edwards, relatifs à la suite de leurs recherches sur l'histoire naturelle du littoral de la France. *Le Temps*. Paris; 1830 November 25: col.5055.

NOTE. Report by Cuvier at the November 15 meeting of the Académie des sciences. A more complete report was published later, rec.504, and another briefer condensation was published in *Le National,* rec.512.

516. [Réponse de M. le Baron Cuvier, Directeur de l'Académie française, au discours de M. de Lamartine, Prononcé dans la Séance du 1er.

avril 1830]. *L'Universel.* Paris; 1830 April 9; (99): 512 and April 11; (101): 519–520.

NOTE. Also published in Académie française, *Discours prononcés dans la séance publique tenue par l'Académie française, Pour la réception de M. de Lamartine et al.*, Paris, Firmin Didot, 1830, p.19–28; in *Le Moniteur universel,* Paris, 1830 April 15; (105): 415–416; in l'Académie française, *Recueil des discours, rapports et pièces* ... in volume for 1830–1839, p.25–36, Paris, Firmin Didot, 1841; and in Paul Souday, *Les romantiques à l'Académie suivi des discours de réception de MM. de Lamartine, et des réponses de MM. le baron Cuvier et al.,* ... Paris, Ernest Flammarion, [1928], p.64–75.

517. ... Note sur quelques ossements qui paraissent appartenir à une espèce, perdue seulement depuis 200 ans. *L'Universel.* Paris; 1830 July 18; (197): 312.

NOTE. Read at the Académie des sciences, July 12, 1830. Also published, but not compared, rec.506. A different reporting of the same paper was also published, rec.507.

518. Notice of some Bones which appear to have belonged to the Dodo, a species of Bird extinct within the last Two Centuries. Read at the Academy of Sciences, 12th July 1830, by Baron Cuvier. *Edinburgh journal of natural and geographical science;* 1830/1831; [3]: 30–32.

NOTE. Translation of rec.506, including the remarks of Blainville and Geoffroy Saint Hilaire, and Cuvier's response to Blainville.

519. [Report on three memoirs of MM. Audouin and Milne Edwards on invertebrates of the coasts of France]. *Edinburgh journal of natural and geographical science;* 1830/1831; 3: 176–177; 243–245.

NOTE. Extracts in English of report given by Cuvier at the November 15, 1830 session of the Académie des sciences. Published more fully, rec.504, from which they were translated.

520. Baron Cuvier's great work on Fishes. *Edinburgh journal of science;* Apr./Oct. 1830; n.s. v.1: 158–162.

NOTE. P.159–162 is an extract which ". . . contains some of the particulars regarding the *Perca fluviatilis,* or common Perch" (cf. p.159), from the *Histoire naturelle des poissons,* Paris, 1828, v.2, p.10–31 in the octavo printing, rec.706, and p.14–23 in the quarto printing, rec.707. The translation is condensed and shortened.

521. On the Mullets of Europe. By Baron Cuvier. *Edinburgh journal of science;* Oct./Apr. 1830; n.s. v.2: 61–66.

NOTE. Condensed translation, "From l'*Histoire naturelle des poissons, . . . ,* v.3 Paris, 1829," p.422–443 in the octavo printing, rec.706 and p.311–326 in the quarto printing, rec.707.

522. Ueber einen Schmarotzerwurm einer neuen Gattung (*Hectocotylus octopodis*) (Vorgelesen der Königl. Akademie der Wissenschaften am 12 Oktober 1829.). *Notizen aus dem Gebiete der Natur= und Heilkunde,* Erfurt; 1830; 27: col.6–10, 4 fig.

NOTE. Translation of rec.459. A different German translation has substantial omissions, rec.552.

523. Rapport sur les *Collections rassemblées au Cap et aux Indes orientales,* par M. Lamarre Picquot. *Académie des sciences. Procès verbaux;* 1921; 9(1828–1831): 629–631.

NOTE. Presented at the meeting of May 9, 1831 of the Académie, by Geoffroy Saint Hilaire, Duméril, and Cuvier as reporter. Condensed and shortened versions, rec.527 and rec.529, were published earlier than the above.

524. Rapport . . . sur le Mémoire de M. Dugès, . . . qui a pour titre *Additions aux recherches sur l'Ostéologie et la Myologie des Batraciens à leurs différents âges. Académie des sciences. Procès verbaux;* 1921; 9(1828–1831): 717–719.

NOTE. Presented at the meeting of Nov. 21, 1831 of the Académie and signed by Duméril, and Cuvier as reporter. An abstracted report was also published, rec.533.

525. Rapport verbal sur la Monographie des Esturgeons de MM. Brandt et Ratzeburg, fait à l'Académie des Sciences. *Annales des sciences naturelles;* 1831; 22: 222–224.

NOTE. Date of session not given.

526. [Rapport de M. le baron Cuvier sur un travail de M. Deshayes ayant pour titre: *Tableaux comparatifs des Coquilles vivantes, avec les Fossiles des terrains tertiaires de l'Europe.* (Fait à l'Académie des Sciences, le 31 octobre 1831)]. *Le Moniteur universel.* Paris; 1831 November 21; (325): 2186–2187.

NOTE. Signed by Ad. Brongniart, Cordier, and Cuvier as reporter. Published also in *Annales des sciences naturelles;* 1831; 24: 176–184; in the *Académie des sciences, Procès verbaux;* 1921; 9(1828–1831): 709–711; and separately, [Paris, Imprimerie de Mme. Ve. Agasse, 1831] 8p. 21cm. A shortened version was published earlier, rec.532. Above was also extracted in rec.530 which was translated into English, rec.550. *Le Moniteur universel* and the separate give Brongniart's first name abbreviation as Alex.

527. [Rapport sur les collections rassemblées au Cap et dans les Indes par M. Lamarre Piquot, . . .]. *Le National.* Paris; 1831 May 13; (133): [2].

NOTE. A condensation of a report read by Cuvier before the Académie des sciences on May 9, 1831. Published more fully, rec.523. Another condensation was also published, rec.529.

528. Rapport sur un mémoire de M. Duvernoy, relatif aux moyens . . . pour distinguer les serpents venimeux. *Le National.* Paris; 1831 May 18; (138): [1–2].

NOTE. Presented at the May 16, 1831 meeting of the Académie des sciences and signed by Duméril, and Cuvier as reporter. Published also in May 1831 in the *Gazette médicale de Paris,* no.21: 183–184; and in the *Bulletin des sciences naturelles;* 1831; 25: 214–218; in the *Académie des sciences, Procès verbaux;* 1921; 9(1828–1831):

632–634; and with some omissions in the *Revue bibliographique;* 1831; 2: 59–61 (with: *Annales des sciences naturelles;* 1831; 23).

529. Collections de M. Lamarepicquot. *Revue bibliographique;* 1831; 2: 57–58 (with: *Annales des sciences naturelles;* 1831; 23).

NOTE. Presented at the May 9, 1831 meeting of the Académie des sciences and signed by Geoffroy Saint Hilaire, Duméril, and Cuvier as reporter. Another condensation was published earlier, rec.527. Published more fully, rec.523.

530. . . . Rapport sur un ouvrage de M. Deshayes, ayant pour titre: *Tableau comparatif des coquilles vivantes avec les fossiles des terrains tertiaires de l'Europe. Revue encyclopédique;* Oct./Dec. 1831; 52: 259–261.

NOTE. Report to the Académie des Sciences made on October 31, 1831 by Cuvier as reporter in his name and those of Brongniart and Cordier. Translated into English, rec.550. The above is extracted from rec.526 with some difference in wording. A large part is same as rec.532.

531. Rapport verbal sur les collections apportées récemment des Indes de M. Dussumier. *Le Temps*. Paris; 1831 January 1; (442): col.5755–5756.

NOTE. Read at the December 13, 1830 session of the Académie des sciences. Greatly condensed version of the report which was published more fully, rec.505.

532. Rapport sur un ouvrage de M. Deshayes ayant pour titre: *Tableaux comparatifs des coquilles vivantes avec les fossiles des terrains tertiaires de l'Europe. Le Temps*. Paris; 1831 November 7; (750): col. 10996–10998.

NOTE. Presented at the session of October 31, 1831, Académie des sciences, in the names also of MM. Brongniart and Cordier with Cuvier as reporter. Published more fully in rec.526. A shortened version of above was also published, rec.530, and translated into English, rec.550.

533. Rapport . . . sur un mémoire de M. Dugez [sur une note supplémentaire relative à l'Ostéologie et à la Myologie des Batraciens (Séance du 21 novembre 1831, Académie des sciences)]. *Le Temps.* Paris; 1831 November 25; (768): col.11283–11286.

NOTE. Also published in *Annales des sciences naturelles;* 1832; 25: 211–214. Above is an abstracted report of the presentation. Published more fully later, rec.524. Signed by Duméril, and Cuvier as reporter.

534. Bericht über den zoologischen Theil der Expedition unter Duperrey. *Isis von Oken,* Jena; 1831; 24: col.1261–1262.

NOTE. Abstract in German of rec.405.

535. G. Cuvier über die Gebeine eines Vogels, der erst seit 200 Jahren ausgestorben zu seyn scheint. *Neues Jahrbuch für Mineralogie, Geologie und Paläontologie,* Stuttgart; 1831: 215–216.

NOTE. Although the above reporting cites rec.507 as its source, it appears to be a translation and condensation of rec.506.

536. Rapport sur un ouvrage de M. Guérin, intitulé: *Iconographie du règne animal.* (Lu à l'Académie royale des sciences, dans la séance du lundi 6 février 1832.). *Annales des sciences naturelles;* 1832; 26: 331–335.

NOTE. Also published separately, [Paris?, J. B. Baillière, 1832], 3p. 21cm. Cuvier's report on Guérin's iconography of his *Règne animal,* rec.802.

537. Mémoire . . . sur les Oeufs de la Seiche; . . . [Lu à l'Académie des Sciences le 2 avril 1832]. *Le Temps.* Paris; 1832 April 4; (899): col.13440–13445.

NOTE. Also published in *Annales des sciences naturelles;* 1832; 26: 69–75. A different reporting of this paper also appeared, rec.546. A summary was published, rec.538. Extracted and summarized in rec.547. Summarized in German, rec.553 and rec.557.

538. Mémoire sur les oeufs de la seiche. *Société philomatique de Paris. Bulletin;* 1832; Ser. IV. v.1(May): 71–73.

NOTE. Report at the April 7, 1832 meeting of the Société philomatique on a report, rec.546, read by Cuvier to the Académie des sciences on April 2, 1832. Same text as rec.546, but above has omissions. Somewhat different reporting appeared earlier, rec.537.

539. Rapport Fait à la Chambre par M. le Baron Cuvier, au nom d'une Commission spéciale chargée de l'examen de dix-huit Projets de loi relatifs à divers emprunts votés par un même nombre de départemens pour l'exécution de la loi du 6 novembre 1831. [n.p.; 183?]. 10p. 8°. (France. Chambre des Pairs. 1814–1848. Impressions diverses. Session de 1831; v. 2, no.93).

NOTE. At head of title: Chambre des Pairs. Séance du 16 janvier 1832.

540. Nouveauté des continens. *La France littéraire;* 1832; 1: 47–55.

NOTE. Extracts from rec.744 (1:241–254), and from another printing, rec.673 (p.176–185). Also published in the *Revue universelle,* Bruxelles, 1832, 1. année, v.1: 170–174. Translated into Hungarian, rec.558.

541. [Sur le projet de loi relatif à la contrainte par corps]. *Le Moniteur universel.* Paris; 1832 January 2; (2): 13.

NOTE. Presented during discussion at the December 31, 1831 meeting of the Chambre des Pairs.

542. [Rapport de la commission chargé de l'examen de 18 projets de loi de l'intérêt local]. *Le Moniteur universel.* Paris; 1832 January 17; (17): 154.

NOTE. Presented at the January 16, 1832 meeting of the Chambre des Pairs by Cuvier as the reporter of the commission.

543. [Defense of Conseil royal in connection with budget of the University], Chambre des Députés. Séance du vendredi 18 [i.e.17?] février 1832. *Le Moniteur universel.* Paris; 1832 February 18; (49): 486.

544. [Rapport de la commission pour le projet de loi sur les céréales]. *Le Moniteur universel.* Paris; 1832 April 11; (102): 1032–1033.

NOTE. Cuvier, as reporter for the commission, presented the above at the April 10, 1832 meeting of the Chambre des Pairs. Also published, but texts were not compared, in: France. Chambre des Pairs, Impressions, N°164 [Session] 1831. Séance du 10 avril 1832. 14 p. with title; *Rapport Fait à la Chambre par M. le Baron Cuvier, au nom d'une Commission spéciale chargée de l'examen du Projet de loi sur les céréales.*

545. Sur le poisson appelé *Machaera.* *Paris. Muséum national d'histoire naturelle. Nouvelles annales;* 1832; 1: 43–49. 1 pl.

NOTE. Read at the March 12, 1832 session of the Académie des sciences. Published also separately, [Paris, Imprimerie de Jules Didot, l'aîné, 1832?], 7p. 1pl. 26cm.

546. Sur les oeufs de Seiche. *Paris. Muséum national d'histoire naturelle. Nouvelles annales;* 1832; 1: 153–160. 1 pl.

NOTE. Read at the April 2, 1832 meeting of the Académie des sciences. An earlier and somewhat different reporting of the same paper was also published, rec.537. Published with omissions, rec.538 and rec.547. Summarized in German, rec.557, and translated into German with omissions, rec.553.

547. Mémoire Sur les oeufs de la seiche. *Revue encyclopédique;* April/June 1832; 54: 202–205.

NOTE. Presented at the session of April 2, 1832. Published more fully, rec.546. Also published with omissions in rec.537, and with omissions and different wording of some sections in rec.538. Translated into German, rec.553.

548. Mémoire ... sur les progrès de l'ossification dans le sternum des oiseaux. [Lu à l'Académie des sciences, séance du 2 janvier 1832.]. *Le Temps.* Paris; 1832 January 4; (808): col.11982–11992.

NOTE. Includes an exchange between MM. Serres and Cuvier. Also published in *Annales des sciences naturelles;* 1832; 25: 260–272 and in the *Gazette médicale de Paris;* 1832; 3(1): 9–11.

549. Rapport sur un mémoire de M. Bennati, intitulé: *Du mécanisme de la voix humaine dans le chant,* p.1–16. In: Bennati, F. Recherches sur le mécanisme de la voix humaine. Ouvrage qui a obtenu un prix à la Société des sciences physiques et chimiques de Paris, précédé du Rapport du MM. G. Cuvier, de Prony et Savart, à l'Académie royale des sciences . . . , Paris: J. B Baillière; 1832: 1–16.

NOTE. Presented at the May 10, 1830 meeting of the Académie des Sciences and signed by de Prony, Savart, and Cuvier as reporter. Published also in the *Académie des sciences, Procès verbaux;* 1921; 9(1828–1831): 441–443. Different condensations were published earlier in newpapers, rec.511.

550. Cuvier and Brongniart's report on M. Deshayes' "Tableau comparatif des coquilles vivantes avec les fossiles des terrains tertiaires de l'Europe." *American journal of science,* New Haven, Conn.; 1832; 1: 196–198.

NOTE. Translation of rec.530 which was extracted from the full report, rec.526.

551. Account of the Common Mackerel, (*Scomber scombrus, Lin.*) and the Garum of the Ancients. By Baron Cuvier and M. Valenciennes. *Edinburgh journal of science;* Oct./Apr. 1832; n.s. v.6: 286–294.

NOTE. Translation of p.16–34 from *Histoire naturelle des poissons,* v.8, 1831 of the octavo printing, rec.706, and p.12–25 of the quarto printing, rec.707.

552. Ueber *Hectocotylus octopodis* von G. Cuvier . . . *Isis von Oken,* Jena; 1832; 25: col.559–562. illus.

NOTE. Condensed translation of rec.459. Lacks 2 of 9 figures from French. Not the same German translation as rec.522, nor is it a translation of the shortened *Le Globe* article, rec.466.

553. Ueber das Ei der Sepie (*Sepia officinalis*). *Notizen aus dem Gebiete der Natur= und Heilkunde,* Erfurt; 1832; 34: col.198–200.

NOTE. Translation of rec.547. Rec.557 is a different and shortened translation into German.

554. Ueber die Organisation der Saamen. (Auszug des Berichtes über die Arbeiten der [franzö.] Akademie der Wissenschaften des Instituts während des Jahres 1829, abgefasst von Herrn Cuvier.). *Justus Liebigs Annalen der Chemie,* Leipzig; 1833; 7: 92–99.

NOTE. Translated extract from rec.649 which is in turn extracted from rec.647.

555. Bericht über die zoologische Ausbeute der *Chevrette* in den indische Meeren. *Isis von Oken,* Jena; 1834; 27: col.1050.

NOTE. Brief German summary of rec.458.

556. Cuviers Bericht über Roulins neuen Tapir in Süd-Amerika. *Isis von Oken,* Jena; 1834; 27: col.1058–1059.

NOTE. German condensation of rec.471.

557. Über das Ey der Dintenschnecke. *Isis von Oken,* Jena; 1835; 28: col.497–499.

NOTE. German summary of rec.537 or rec.546 A more complete and a different German translation was also published, rec.553.

558. A' szárazföldek' újsága. Cuvier' nyomozásai. *Tudománytár;* 1835; 7: 92–100.

NOTE. Hungarian translation of rec.540.

559. Krótki rzut oka Na funkcye czyli czynności umysłowe zwierząt. (Z Cuvier). [Brief look at the intellectual functions or activities of animals]. *Biblioteka Warszawska;* 1841; 1: 382–387.

NOTE. Polish translation and apparent condensation of *Règne animal,* 1817, rec.747, v.1: 47–55; 1829, rec.748, v.1: 40–46.

560. Wyjątki z historyi nauk przyrodzonych p. Jerzego Kiuwiera (Cuvier). Przekładu Gustawa Belkego i Alexandra Kremera. *Biblioteka Warszawska;* 1852; 2: 479–510.

NOTE. Polish translation by Gustaw Belke and Aleksander Kremer of extracts from v.1, lessons 6 and 8; v.2, lessons 1 and 9; v.3, lesson 11, of *Histoire des sciences naturelles . . .* , rec.702. The latter was translated into Polish, rec.704.

561. [Développement physique et moral de l'homme]. In: Barasin, Juliu. Cursulú elementariú de istoria naturală pentru usulú classiloru superióre gimnasiale în România. Tomulu I Zoologia compussĕ de Dr. Juliu Barasin, . . . [Translation: Elementary course of natural history for the use of senior high school classes in Romania. v.1; Compound zoology]. Bucurescĭ: Imprimeria Statuluĭ; 1864; 1: 196–198.

NOTE. The information on the above was supplied by Dr. M. Bacescu, Musée d'Histoire Naturelle, Bucarest, Romania. The work cited above was not located in this country nor in those libraries visited in Europe, nor was it listed in those printed catalogues, bibliographies consulted, etc. It is a translation from *Le Règne animal,* probably the Paris, 1829 edition, rec.748, v.1, p.75–80, although it has not been compared. The same extract was also translated into Polish, rec.334, but it was not compared with the above.

562. De la part à faire aux sciences et aux lettres dans l'instruction publique. Extrait d'un article du "Moniteur universel" du mardi 3 novembre 1807 par Cuvier. *Revue internationale de l'enseignement;* 1885; 10: 380–387.

NOTE. At head of title of article: Revue rétrospective des ouvrages de l'enseignement. Also published in 1807, rec.217. Included in the above, but not in rec.217, the earlier publication, is a letter from Cuvier to M. Lecoz, Archevêque de Besançon, Dec. 8, 1807, containing an analysis of A. Brongniart's *Traité élémentaire de minéralogie avec des applications aux arts . . . .*

563. Krass, M. Ein französischer Bericht über das Schulwesen in Niederdeutschland aus dem Jahre 1811. . . *Gesellschaft für deutsche Erziehungs- und Schulgeschichte. Mitteilungen;* 1895; V(Heft 2): 123–137.

NOTE. Includes translated excerpts from rec.731 quoted by Krass in his article.

564. Rapport à l'Assemblée des professeurs du Muséum d'histoire naturelle. In: Hamy, E. T., ed. Dicquemare jugé par Cuvier (1805) . . . Paris. Muséum national d'histoire naturelle. Bulletin; 1906; 12: 183–184.

NOTE. Report by Lamarck and Cuvier, as reporter, dated Sept. 21, 1805 and read before the Assembly Sept. 25, 1805 in response to a request for advice about publication of the Abbé Dicquemare's portfolio consisting mostly of memoirs on natural history.

565. Note instructive sur les recherches à faire relativement aux différences anatomiques des diverses races d'hommes. *Revue de l'École d'Anthropologie de Paris;* 1910; 20: 303–306.

NOTE. Also published in *Société des Observateurs de l'Homme en l'an VIII,* published and presented by Jean Copans and Jean Jamin, Paris, Le Sycamore, [1978], p.173–176.

566. Osservazioni sul cadavere di una donna nota a Parigi e a Londra con il nome di "Venere ottentotta" . . . di Georges Cuvier. *Kos;* 1985; 2(13): 79–80.

NOTE. Translation of rec.328 by Simone Piatelli with a few indicated omissions.

ANNUAL REVIEW OF SCIENCE

567. Notice de la partie physique des travaux de la classe des sciences physiques et mathématiques, pendant le dernier trimestre de l'an VII; lue à la séance publique du 15 vendémiaire an VIII [Oct.7, 1799]. *Magasin encyclopédique;* 1799; 5. année(4): 223–239.

NOTE. Covers period June 19–Sept. 22, 1799. Published also in the *Décade philosophique;* 1799; [v.23] (6): 321–333. Extracted in English, rec.568.

568. [. . . Account of the labours of the Mathematical and Physical Class during the last quarter of the year 7 . . .]. *Philosophical magazine,* London; [1800]; 6: 84–88.

NOTE. Read at the October 7, 1799 sitting of the French National Institute, covering June 19 to Sept. 22, 1799. Translated extracts from rec.567.

569. Compte rendu des travaux de la classe des sciences physiques et mathématiques de l'Institut national, Partie physique, pendant les mois de vendémiaire, brumaire et frimaire an 8, lu à la séance publique du 15 nivôse [Jan. 5, 1800]. Partie physique. *Magasin encyclopédique;* 1799; 5. année(5): 244–256.

NOTE. Covers period Sept. 23–Dec. 21, 1799. Also published in the *Décade philosophique;* 1800; [v.24] (12): 129–138, and extracted in English, rec.570.

570. [Account of the labours of the Physical and Mathematical class during the preceding three months]. *Philosophical magazine,* London; [Feb./May, 1800]; 6: 184–189.

NOTE. Read at the January 5, 1800 sitting of the Institut National. English extracts from rec.569.

571. Notice des travaux de la classe des sciences mathématiques et physiques de l'Institut national, pendant le deuxième trimestre de l'an 8: lue à la séance publique du 15 germinal [April 5, 1800]. Partie physique. *Magasin encyclopédique;* 1800; 6. année(1): 89–103.

NOTE. Covers period Dec. 22, 1799–March 21, 1800. Published also in *Décade philosophique;* 1800; [v.25] (21): 129–140; and extracts were translated into English, rec.572.

572. [Account of the labours of the Mathematical and Physical Class during the preceding three months]. *Philosophical magazine,* London; [Feb./May, 1800]; 6: 275–282.

NOTE. Presented at the April 5, 1800 sitting. Translated extracts from rec.571.

573. Notice des travaux de la classe des sciences mathématiques et physiques de l'Institut national, pendant les mois de germinal, floréal et prairial an 8: lue à la séance publique du 15 messidor [July 4, 1800]. Partie physique. *Magasin encyclopédique;* 1800; 6. année(2): 375–390.

NOTE. Covers period March 22–June 19, 1800. Published also in the *Décade philosophique;* 1800; [v.26] (31): 193–203. Extracts translated into English, rec.574.

574. [Account of the labours of the Mathematical and Physical Class during the preceding three months]. *Philosophical magazine,* London; [June/Sept., 1800]; 7: 368–373.

NOTE. Read at the July 4, 1800 sitting of the Institut de France. Translated excerpts of rec.573.

575. Notice des travaux de la classe des sciences physiques et mathématiques de l'Institut national, du 15 messidor an 8 jusqu'au 15 ven-

démiaire an 9. Partie physique. *Magasin encyclopédique;* 1800; 6. année(4): 93–97.

NOTE. No author is given for this, although Delambre is credited for the Partie mathématique. Covers period July 4, 1800–October 7, 1800.

576. Notice des travaux de la Classe des sciences mathématiques et physiques, pendant le second trimestre de l'an 9. Partie physique. *Magasin encyclopédique;* 1801; 6. année(6): 371–388.

NOTE. Covers period Dec. 22, 1800–March 21, 1801. Published also as a separate, with the Partie mathématique by Delambre. [Paris, Baudouin, 1801, p.16–32, 4°]; and extracted and condensed, rec.92 and rec.577.

577. [Notice des travaux de la classe des sciences mathématiques et physiques, pendant le second trimestre de l'an 9]. Partie physique. *Décade philosophique;* 1801; [29](21): 130–139.

NOTE. Extracts and condensation of rec.576.

578. Notice des travaux de la Classe des sciences mathématiques et physiques, pendant le troisième trimestre de l'an 9. Partie physique. *Magasin encyclopédique;* 1801; 7. année(2): 94–110.

NOTE. Deals with period March 22–June 19, 1801. Also published as *Partie physique* in: "Notice des travaux de la classe des sciences mathématiques et physiques, Pendant le troisième trimestre de l'an 9," p.16–32. [Paris, Baudouin, Imprimeur, n.d.]. Published with omissions, rec.579 and rec.580.

579. Extrait de la Notice des travaux de la classe des sciences mathématiques et physiques, pendant le troisième trimestre de l'an 9. (Partie Physique). *Décade philosophique;* 1801; [30](30): 129–138.

NOTE. Extracts from rec.578, but not the same as rec.580.

580. Notice des travaux de la classe des sciences mathématiques et physiques de l'Institut de France, pendant le troisième trimestre de l'an 9. *Journal de physique;* 1801; 53: 76–81.

NOTE. Extracts from rec.578, but not the same abstracted publication as rec.579.

581. [Notice des travaux de la Classe des sciences mathématiques et physiques, pendant le dernier trimestre de l'an 9]. Partie physique. *Magasin encyclopédique;* 1801; 7. année(3): 393–415.

NOTE. Covers period June 20–Sept. 22, 1801. Also published in the *Décade philosophique;* 1801; [31](4): 193–201; (5): 257–264; and in: ["Notice des travaux de la classe des sciences mathématiques et physiques, Pendant le dernier trimestre de l'an 9"], *Partie physique;* p.17–39, [1801]. Published with an omission, rec.582. Translated into English, rec.583.

582. Partie physique [Notice des travaux de la classe des sciences mathématiques et physiques, pendant le dernier trimestre de l'an 9]. *Journal de physique;* 1801; 53: 389–402.

NOTE. Covers period June 20–Sept. 22, 1801. With exception of the omission of one substantial paragraph in the above, text is same as that of rec.581.

583. [French National Institute. Notice of the labours of the Class of the Mathematical and Physical Sciences during the last three months of the year 9]. [June 20–Sept. 22, 1801]. Physics. *Philosophical magazine,* London; [1801/1802]; 11: 187–188, 271–280.

NOTE. Translation of rec.581.

584. Compte Rendu des travaux de la Classe des sciences mathématiques et physiques de l'Institut national, Depuis le premier messidor an 12, jusqu'à pareil jour de l'an 13. Partie physique. *Journal de physique;* 1805; 61: 63–77.

NOTE. Covers June 20, 1804–June 20, 1805. Also published separately, [Paris, Baudouin, 1805?]: 22p. 4°. Translated into English with exception of section on agriculture, rec.585; and into German in a shortened version, rec.586.

585. An account of the Labours of the Class of Mathematics and Physical Sciences of the French National Institute from the 20th of June 1804 to the same Day, 1805. By M. Cuvier, . . . Physical Part. *Philosophical magazine,* London; [June/Sept., 1805]; 22: 172–178, 274–278, 354–355.

NOTE. Translation of rec.584, with exception of section on agriculture.

586. Nachricht von den Arbeiten der physisch=mathematischen Classe des National=Instituts vom I. Messidor 12. bis dahin 13. (Vom Hrn. Cuvier, . . .). *Magazin für den neuesten Zustand der Naturkunde,* Jena; 1805; 10: 236–254.

NOTE. Translation and condensation of rec.584.

587. Analyse Des travaux de la Classe des Sciences Mathématiques et Physiques de l'Institut national, depuis le Ier. messidor an 13 [20 juin 1805], jusqu'au Ier. juillet 1806; . . . Lue à la Séance publique du 17 juillet. Partie physique. *Journal de physique;* 1806; 63: 156–165, 169–183.

NOTE. Also published in the *Magasin encyclopédique;* 1806; 4: 405–437; and in *Académie des sciences. Mémoires;* 1806; [7]; (1)(Hist.): 43–79. Translated into German, rec.589; and into English with omissions, rec.588.

588. [Analysis of labours of the class of physical and mathematical sciences, from 20th of July 1805 to the first of July 1806]. *Philosophical magazine,* London; October, 1806–May, 1807; 26: 183–187, 272–275, 368–373; 27: 87–90, 183–188.

NOTE. Translation of rec.587 with omissions.

589. Nachricht von den Arbeiten der mathematisch=physischen Classe des National Instituts seit dem 20. Jun. 1805 bis zum I. Jul. 1806. Physischer Theil. Vom Hrn. Cuvier, . . . (Vorgelesen am 7. Jul. in der öffentlichen Sitzung). *Magazin für den neuesten Zustand der Naturkunde,* Jena; 1806; 12: 325–377.

NOTE. Translation of rec.587 with small omission.

590. [Analyse des travaux de la classe des sciences mathématiques et physiques de l'Institut national, pendant le deuxième semestre de 1806. Partie physique]. . . . *Le Moniteur universel.* Paris; 1807 January 11; (11): 41–42; January 14; (14): 53–54; January 16; (16): 60–61; January 17; (17): 65–66.

NOTE. Read at the meeting of January 5, 1807. Also published in *Magasin encyclopédique;* 1807; (1): 421–452 and separately, [Paris, Baudouin, Imprimeur, 1807?]: 38p. 4°; and in *Académie des sciences. Mémoires;* 1806; [7] (1)(Hist.): 80–116. Translated into English with omissions, rec.591.

591. Proceedings of the [French National] Institute for the last half-year of 1806. Analysis of the Labours in the Physical Department, as drawn up by M. Cuvier. *Philosophical magazine,* London; 1807; 28: 280–284, 366–372; 29: 85–89, 184–187, 277–284.

NOTE. Translation of rec.590 with exception of section on Cuvier's work and part of section on Humboldt.

592. Analyse des travaux de la Classe des sciences mathématiques et physiques de l'Institut, pendant l'année 1807. Partie physique. *Le Moniteur universel.* Paris; 1808; February 20; (51): 205–206; February 21; (52): 208–209; February 22; (53): 212–213.

NOTE. Also published in *Magasin encyclopédique;* 1808; (1): 178–210 and separately, [Paris, Baudouin, 1808?], 36p. 4°; and in *Académie des sciences. Mémoires;* 1807; [8] (1)(Hist.): 58–92. Translated and condensed in English, rec.593.

593. [Analysis of the Labours of the Class of Mathematical and Physical Sciences of the French Institute, for the Year 1807]. *Philosophical magazine,* London; 1809; 34(136): 144–155.

NOTE. Translation, condensed and with omissions, of rec.592.

594. Analyse des travaux de la classe des sciences mathématiques et physiques de l'Institut pendant l'année 1808. Partie physique. *Le Moniteur universel.* Paris; 1808 [i.e.1809]; January 5; (5): 16–18; January 6; (6): 20–22.

NOTE. Distributed at the public meeting of the Class on Jan. 2, 1809. Also published in *Magasin encyclopédique;* 1809; (1): 152–178 and in 1809 in *Académie des sciences. Mémoires;* 1808; [9] (1)(Hist.): 53–84 and separately, [Paris, Baudouin, imprimeur; 1809?] 34p. 4°. Extracts were translated into English with some condensation, rec.595 and rec.596.

595. [Analysis of the work of the *Académie des sciences, Classe des sciences mathématiques et physiques, Partie physique* during the year 1808]. *Journal of natural philosophy;* 1809 Sept./Oct.; 24: 70–79.

NOTE. Translation of extracts from rec.594 with some condensation, but not the same as rec.596.

596. [Proceedings of the French National Institute for the year 1808]. Physics. *Philosophical magazine,* London; 1809; 34: 234–237, 302–314.

NOTE. Extracts and some condensation of rec.594, but not the same as rec.595.

597. Analyse des travaux de la classe des sciences mathématiques et physiques de l'Institut, pendant 1809. *Le Moniteur universel.* Paris; 1810 January 9; (9): 33–34; January 10; (10): 37–38; January 12; (12): 45–46; January 13; (13): 48–50.

NOTE. Distributed to the public session of the Classe des sciences . . . January 2, 1810. *Le Moniteur universel* does not attribute to Cuvier, nor to anyone. See Note to rec.598 about authorship. Published with omissions, rec.599. Another publication of this review (rec.598) was not compared with the above. Translated into English, rec.601, but not compared with above.

598. Analyse des travaux de la classe des sciences mathématiques et physiques de l'Institut, pendant l'année 1809. Partie physique. [Paris: Baudouin; 1810?]. 37p. 4°.

NOTE. "Distribué à la séance publique de la Classe, le 2 janvier 1810." There is some doubt about the authorship of this report. It may be by Delambre. Cuvier's name does not appear, but he is credited with it in the index of *Académie des sciences. Mémoires,* rec.599, and in the Bibliothèque Nationale catalogue. The *Magasin encyclopédique,* rec.600, credits Delambre with the full report including the Partie physique.

Appeared in *Le Moniteur universel,* rec.597, but the latter was not compared with above. *Le Moniteur universel* does not indicate authorship. Published with omissions in rec.599 and with different omissions in rec.600. Translated into English with some minor changes, rec.601.

599. Analyse Des travaux de la classe des sciences mathématiques et physiques de l'Institut, pendant l'année 1809. Partie physique. *Académie des sciences. Mémoires;* 1809; (1)(Hist): 51–95.

NOTE. Published in 1810. Another publication of rec.597 and rec.598 with omissions. Published also with different omissions, rec.600. There is some doubt about the authorship of this paper, although the table of contents of the volume gives Cuvier as author, but does not give his name following the title as was customary. See Note for rec.598 for further discussion of the authorship.

600. Analyse des travaux de la classe des sciences mathématiques et physiques de l'Institut, pendant l'année 1809. Partie physique. *Magasin encyclopédique;* 1810; (1): 111–152.

NOTE. Another publication with omissions of rec.598, but not the same as rec.599. There is some doubt about the authorship of this paper. See Note for rec.598.

601. Analysis of the Labours of the Class of Mathematical and Physical Sciences of the French Institute, for the Year 1809. *Philosophical magazine,* London; Jan./June 1810; 35: 306–317, 386–399, 465–466.

NOTE. Translation with some omissions of rec.598. Not compared with rec.597. There is some doubt about the authorship of this report. See Note for rec.598.

"Translated from the original, distributed at the public sitting of the class 2d Jan. 1810." Cf. footnote, p.306 of above.

602. Analyse des travaux de la Classe des sciences mathématiques et physiques de l'Institut, pendant l'année 1810. Partie physique. *Le Moniteur universel,* Paris; 1811 January 12; (12): 45–46; January 15; (15): 58; January 17; (17): 66; January 18; (18): 69–70; January 20; 20: 79–80.

NOTE. Also published in *Magasin encyclopédique,* 1811; 1; 91–54 and in *Académie des sciences. Mémoires;* 1810; 11 (1)(Hist.): xlv–xciij (published in 1811); and separately, [Paris: Baudouin, Imprimeur; n.d.]; 55p. 4°. Selections were translated into English and partially condensed, rec.603.

603. [A brief abstract of the Report of the National Institute at Paris for the year 1810]. *Philosophical magazine,* London; 1811; 37: 154–157.

NOTE. Translation of condensed abstract from rec.602.

604. Analyse des travaux de la Classe des sciences mathématiques et physiques de l'Institut impérial, pendant l'année 1811; Partie physique. *Le Moniteur universel.* Paris; 1812 January 17; (17): 66–68; January 21; (21): 81–82; January 29; (29): 115–116.

NOTE. Also published in *Magasin encyclopédique;* 1812; 1: 308–356, and separately, [Paris, Imprimerie de Firmin Didot, n.d.], 38p. 4° and in *Académie des sciences. Mémoires;* 1811; 12 (1)(Hist.): lxxix–cxiv, published in 1814. Extracts were translated into English, rec.605.

605. [Report of the proceedings of the Imperial Institute of France during the year 1811]. *Philosophical magazine,* London; July, 1812/June, 1813; 40: 462–468; 41: 62–71.

NOTE. Translated extracts of rec.604.

606. Analyse des travaux de la classe des sciences mathématiques et physiques de l'Institut impérial, pendant l'année 1812. Partie physique. *Le Moniteur universel.* Paris; 1813 January 21; (21): 81–82; February 22; (53): 208–210; February 24; (53): 217–218; February 26; (57): 223–224.

NOTE. Published also with omission of two small paragraphs in the *Magasin encyclopédique;* 1813 (1): 241–310; in *Académie des sci-*

ences. Mémoires; 1813; [13] (1)(Hist.): lxxxj–cxxxij, published in 1814; and separately, [Paris, Imprimerie de Firmin Didot; 1813]; 47p. 4°. Extracts were translated into English, rec.607.

607. Imperial Institute of France for the year 1812, drawn up by M. Cuvier. *Philosophical magazine,* London; 1813; 41: 315–316, 380–392; and 42: 74–76, 142–145.

NOTE. Translated extracts from rec.606 with an apparent insertion.

608. Analyse des travaux de la Classe des sciences mathématiques et physiques de l'Institut impérial de France, pendant l'année 1813. Partie physique. *Le Moniteur universel.* Paris; 1814 January 20; (20): 78–79; January 21; (21): 83–84; January 23; (23): 91–92; January 26; (26): 102–103; January 27; (27): 107; January 29; (29): 114–115; February 2; (33): 129–130; January [i.e. February] 3; (34): 134–135.

NOTE. Published also separately, [Paris: Imprimerie de Firmin Didot; 1814]; 46p. 4°; and in 1818 in *Académie des sciences. Mémoires;* 1813–1815; [14] (1)(Hist.): cxvij–clxj. Translated into English, rec.609.

609. Proceedings of the French Institute for the year 1813. By M. Cuvier. *Philosophical magazine,* London; 1814 Jan./ June; 43: 458–461; July/Aug., 44: 50–63; 140–149.

NOTE. Translation of rec.608 with a few omissions.

610. [Analyse Des Travaux de l'Académie Royale des Sciences, pendant l'année 1814. Partie physique]. *Le Moniteur universel.* Paris; 1815 March 22; (81): 329–330; March 23; (82): 334; June 2; (153): 621–622; July 5; (186): 768; July 8; (189): 777–778.

NOTE. Also published in 1818 in *Académie des sciences. Mémoires;* 1813–1815; [14] (1)(Hist.): clxij–cxcv, with very slight differences. The latter was also published separately, [Paris: Imprimerie de Firmin Didot; 181?]; 35p. 4°. Extracts were translated into English, rec.611.

611. Abstract of the Labours of the Class of Mathematical and Physical Sciences during the Year 1814. By M. Cuvier, . . . *Philosophical magazine,* London; Jan./June, 1815; 45: 228–232; 310–313.

NOTE. Translated extracts from rec.610.

612. [Analyse Des Travaux de l'Académie Royale des Sciences, pour les années 1813, 1814, 1815. Partie physique]. Année 1815. *Académie des sciences. Mémoires;* 1813–1815; [14] (1)(Hist.): cxcv–ccxxxvj.

NOTE. Published in 1818. Also published separately, [Paris: Imprimerie de Firmin Didot; 181?]; 48p. 4°. Translated into German with omissions and additions, rec.614. Extracts were translated into English, rec.613. Extracts with some condensation were also translated into Italian, rec.615. Cuvier starts this review with a paragraph lamenting the horrors of wars and on the consolation and calmness of science in the midst of turmoil. This paragraph was also published without his following yearly review in *Le Moniteur universel;* 1816 January 10; (10): 36–37.

613. Analysis of the Labours of the Class of Mathematical and Physical Sciences for the Year 1815. By M. Cuvier. *Philosophical magazine,* London; Jan./June, 1816; 47: 311–312; 390–394; 459–465.

NOTE. Extracts from rec.612.

614. [*Jahresbericht über die Arbeiten der Franzosen für 1815*]. Physikalische Wissenschaften. *Isis von Oken,* Jena; 1817; 1: col.399–426.

NOTE. Translation of rec.612 with omissions and additions, and with editorial comments [by L. Oken?] included in text.

615. Analisi compendiata dei lavori della classe di Scienze fisiche e matematiche dell' Istituto Reale di Francia dell' anno 1815, del sig. cav. Cuvier, . . . Parte fisica. *Biblioteca italiana,* Milan; 1816; 2(April/June): 90–104; 249–253.

NOTE. Translated extracts and some condensation of rec.612.

616. Analyse des travaux de l'Académie pendant l'année 1816. *Annales encyclopédiques;* 1817(2): 62–96.

NOTE. Also published separately, [Paris: Imprimerie de Firmin Didot; n.d.]; 32p. 4°, and in *Académie des sciences. Mémoires;* 1816; [Ser. II] 1(Hist.): cxiii–cxliv, published in 1818. Translated into German, rec.618, and extracted in English, rec.617. Extracts dealing with Martinique were also published, rec.323.

617. Analysis of the Labours of the Royal Academy of Sciences of the Institute of France during the Year 1816. Physical Part. By M. Cuvier, . . . *Philosophical magazine,* London; Jan./June, 1817; 49: 329–344.

NOTE. Translated extracts of rec.616.

618. Bericht über die Arbeiten des Königl. französischen Academie der Wissenschaften im Jahr 1816. Von Cuvier. *Isis von Oken,* Jena; 1818: col.3–22.

NOTE. Translation, with editorial comments in text [by L. Oken?], of rec.616. Comments are indicated by brackets.

619. Analyse des travaux de l'Académie royale des sciences, pendant l'année 1817. Partie physique. *Académie des sciences. Mémoires;* 1817; [Ser.II] 2(Hist.): xciij–clxix.

NOTE. Published in 1819. Also published separately; [Paris: Imprimerie de Firmin Didot; n.d.]; 50p. 4°. Extracted with some condensation in 1818, rec.620. Judging from dates of publication, the extract would appear to be from the separate or from possible access to the manuscript.

620. [Compte des travaux de l'Académie relatifs à la physique et à l'histoire naturelle pendant l'année 1817]. *Annales encyclopédiques;* 1818(2): 299–317.

NOTE. Extracts with some condensation of rec.619.

621. Analyse des Travaux de l'Académie Royale des Sciences, pendant l'année 1818. Partie physique. *Académie des sciences. Mémoires;* 1818; [Ser. II], 3(Hist.): clxxix–ccxxx.

NOTE. Published in 1820. Also published separately, [n.p., n.d.], 50p. 4°. Extracted in rec.622.

622. [Analyse des travaux de l'Académie royale des sciences de l'Institut de France, pendant l'année 1818]. . . . Partie physique, . . . Statistique. *Annales maritimes et coloniales;* 1819; Pt.2, [10]: 579–582.

NOTE. Extracts from rec.621, reporting the work of Moreau de Jonnès.

623. Analyse des travaux de l'Académie royale des sciences, pendant l'année 1819. Partie physique. *Annales générales des sciences physiques,* Bruxelles; 1820; 4: 1–24.

NOTE. Also published in 1824 in *Académie des Sciences. Mémoires;* 1819–1820; Ser. II, 4 (Hist.): lxxxj–cxxvj; and separately, [n.p., 182?], 46p. 4°. Extracted in rec.624, rec.361, and rec.433.

624. [Analyse des travaux de l'Académie royale des sciences de l'Institut de France, pendant l'année 1819, en ce qui intéresse la marine et les colonies]. Partie physique. Zoologie, physiologie animale et anatomie. *Annales maritimes et coloniales;* 1820; Pt.2, [12]: 369–372.

NOTE. Extracts from zoology, physiology, and anatomy sections of rec.623 reporting the work of Moreau de Jonnès on animals of the Antilles.

625. Analyse des travaux de l'Académie royale des sciences pendant l'année 1820. Partie physique. *Académie des sciences. Mémoires;* 1819–1820; [Ser.II] 4(Hist.): ccxxix–cccxvij.

NOTE. Published in 1824. Also published separately, [n.p.; 182?], 89p. 4°; and extracted, rec.626, rec.361, and rec.433.

626. [Analyse des travaux de l'Académie royale des sciences de l'Institut de France, pendant l'année 1820, en ce qui intéresse la marine et les colonies]. Partie physique. *Annales maritimes et coloniales;* 1821; Pt.2, [14]: 994–998.

NOTE. Extracts from the meteorology, zoology, and medical sections of rec.625, chiefly on the work of Moreau de Jonnès.

627. Analyse des Travaux de l'Académie Royale des Sciences, pendant l'année 1821. Partie physique. *Académie des sciences. Mémoires;* 1821–1822; [Ser.II] 5(Hist.): 153–203.

NOTE. Published in 1826. Also published separately; [n.p., 182?]; 49p. 4°. Extracted in rec.628.

628. [Analyse des travaux de l'Académie royale des sciences de l'Institut de France, pendant l'année 1821, en ce qui intéresse la marine et les colonies]. Partie physique. *Annales maritimes et coloniales;* 1822; Pt.2, [17]: 313–318.

NOTE. Extracted from zoology and meteorology sections of rec.627.

629. Analyse des Travaux de l'Académie Royale des Sciences, pendant l'année 1822. Partie Physique. *Académie des sciences. Mémoires;* 1821–1822; [Ser.II] 5(Hist.): 321–382.

NOTE. Published in 1826. Also published separately; [Paris, Imprimerie de Firmin Didot, 182?]; 62p. 4°. Extracted in rec.630.

630. [Analyse des travaux de l'Académie royale des sciences de l'Institut de France, pendant l'année 1822, en ce qui concerne la marine et les colonies]. Partie physique. *Annales maritimes et coloniales;* 1823; Pt.2, [19]: 531–535.

NOTE. Extracts from rec.629.

631. Analyse des Travaux de l'Académic Royale des Sciences, pendant l'année 1823. Partie physique. *Académie des sciences. Mémoires;* 1823; [Ser.II] 6(Hist.): lxxxiij–clix.

NOTE. Published in 1827. Also published separately; [Paris: Imprimerie de Firmin Didot; n.d.]; 77p. 4°. Extracts of various sections were also published, rec.398, rec.399, rec.632, rec.633.

632. Analyse des travaux de l'Académie royale des sciences, pendant l'année 1823. . . (Physiologie et anatomie comparée du systeme nerveux). *Annales des sciences naturelles;* 1824; 2: 423–443.

NOTE. Extracts from rec.631.

633. [Analyse, en ce qui concerne la marine et les colonies, des travaux de l'Académie royale des sciences, pendant l'année 1823]. Partie physique. *Annales maritimes et coloniales;* 1824; Pt.2, [18](2): 189–194.

NOTE. Extracts from the plant physiology, zoology, and surgery sections of rec.631.

634. Analyse des Travaux de l'Académie Royale des Sciences, pendant l'année 1824. Partie physique. *Académie des sciences. Mémoires;* 1827; Ser.II 7(Hist.): cxj–cxciv.

NOTE. Published in 1831. Also published separately, [Paris, Imprimerie de Firmin Didot, 182?]; 84p. 4°. Extracted in rec.406, rec.635, and rec.636 which was summarized in German, rec. 637.

635. Analyse des travaux physiologiques de l'Académie royale des Sciences pendant l'année 1824. *Annales des sciences naturelles;* 1825; 5: 225–235.

NOTE. "Physiologie" section of rec.634. Summarized in German, rec.637.

636. [Analyse, en ce qui concerne la Marine et les Colonies, des Travaux de l'Académie royale des Sciences, pendant l'année 1824]. Partie physique. *Annales maritimes et coloniales;* 1825; 2e. Ptie.[26] (2): 149–156.

NOTE. Extracts from the volcanology, zoology, and medicine sections of rec.634, chiefly referring to the work of Moreau de Jonnès.

637. G. Cuvier, Analyse der physiologischen Arbeiten der Academie während 1824. *Isis von Oken,* Jena; 1831; 24: col.800–801 [i.e.780–781].

NOTE. A very brief German summary of rec.635.

638. Analyse des Travaux de l'Académie royale des sciences, pendant l'année 1825. Partie physique. *Académie des sciences. Mémoires;* 1829; [Ser.II] 8(Hist.): lxxxix–cxliv [i.e. cxliij].

NOTE. Also published separately, [Paris, Imprimerie de Firmin Didot, 182?]; 55p. 4°.

639. Analyse Des Travaux de l'Académie royale des Sciences, pendant l'année 1826. Partie Physique. *Académie des sciences. Mémoires;* 1830; [Ser.II] 9(Hist.): xcvij–clxvij.

NOTE. Also published separately; [Paris, Imprimerie de Firmin Didot, 182?]; 71p. 4°. "Distribué à la séance publique du 11 juin 1827."; cf. p.71 of the separate. Extracted in English, rec.640.

640. Extracts from the analysis of the labours of the Academy of Sciences during the year 1826; by Baron Cuvier. *Edinburgh journal of natural and geographical science;* 1830; 2: 378–380, 457–458; 1830/1831; [3]: 113–114.

NOTE. Translation of the zoology section of rec.639.

641. Analyse Des Travaux de l'Académie royale des Sciences, pendant l'année 1827. Partie Physique. *Académie des sciences. Mémoires;* 1831; [Ser.II] 10(Hist.): ciij–cxc.

NOTE. Also published separately; [Paris, Imprimerie de A. Firmin Didot; n.d.]; 88p. 4°; and with omissions, rec.642. Excerpted in rec.432.

642. Extrait de l'analyse des travaux de l'Académie royale des Sciences, pendant l'année 1827, pour la minéralogie et la géologie, la physiologie végétale et la botanique, la zoologie, l'anatomie et la physiologie animale. *Annales des sciences naturelles;* 1828; 14: 300–339 [i.e.355].

NOTE. Published more fully in 1831, rec.641.

643. Analyse Des Travaux de l'Académie royale des Sciences, pendant l'année 1828. Partie physique. *Académie des sciences. Mémoires;* 1832; [Ser.II] 11(Hist.): cxix–ccxxxj.

NOTE. Also published separately; [Paris, Imprimerie de A. Firmin Didot, 183?]; 113p. 4°. Extracts were published in 1829, rec.644, and were translated into English with some omissions, rec.645. Extracts were translated into German, rec.646.

644. Extrait de l'Analyse des travaux de l'Académie royale des Sciences, pendant l'année 1828. *Annales des sciences naturelles;* 1829; 17: 319–365.

NOTE. Extracts from Cuvier's annual "Analyse des travaux . . . ," published in full later, rec.643. Some of these extracts were translated into English, rec.645, with omissions.

645. Extract from the Analysis of the Labours of the Academy of Sciences during the year 1828. By Baron Cuvier. *Edinburgh journal of natural and geographical science;* 1830; 1: 144–146, 214–218, 293–297.

NOTE. Translated with omissions from rec.644.

646. Über die Arbeiten der kön. Akademie der Wissenschaften im Jahr 1828. *Isis von Oken,* Jena; 1834: 1082–1086.

NOTE. Translation of extracts from rec.643. Not compared.

647. Analyse des travaux de l'Académie royale des sciences, pendant l'année 1829. Partie physique. [Paris: Imprimerie de A. Firmin Didot; 18?]. 137p. 4°.

NOTE. Extracts were published in rec.648 and extracts of medical interest in rec.649. One small section of the latter was translated into German, rec.554. Apparently the above was not published in *Académie des sciences. Mémoires,* as were the earlier *Analyses. . . .*

648. Extrait de l'Analyse des travaux de l'Académie royale des Sciences pendant l'année 1829. *Annales des sciences naturelles;* 1830; 21: 209–281.

NOTE. Extracts from rec.647. Some are the same extracts as those in rec.649.

649. Extrait du compte-rendu des travaux de l'Académie des Sciences de l'Institut, pendant l'année 1829, rédigé par M. Cuvier, en ce qui concerne

la médecine et les sciences accessoires à la médecine. *Transactions médicales;* 1830; 1(August): 260–277, 409–425.

NOTE. Extracts of medical interest from rec.647. Not all are the same extracts as in rec.648. An extract from the botanical section on the structure of seed was translated into German, rec.554.

650. Analyse des Travaux de l'Académie royale des Sciences, pendant l'année 1830. Partie physique. *Académie des sciences. Mémoires;* 1838; [Ser.II] 16(Hist.): A,i–cxvij.

NOTE. Published separately, [n.p., n.d.]: 117p. 4°, and with omissions, rec.651.

651. Analyse des travaux de l'Académie royale des Sciences, pendant l'année 1830. *Annales des sciences naturelles;* 1831; 24: 184–239.

NOTE. A more complete publication appeared in 1838, rec.650.

652. Analyse des Travaux de l'Académie Royale des Sciences pendant l'année 1831. Partie Physique. *Académie des sciences. Mémoires;* 1838; Ser.II 16(Hist.): cxix–ccxxxix.

MONOGRAPHS

653. Anatomie comparée. Recueil de planches de myologie dessinées par Georges Cuvier ou exécutées sous ses yeux par M. Laurillard. Publié sous les auspices de M. le Ministre de l'instruction publique et sous la direction de MM. Laurillard . . . , et Mercier . . . Paris: Chez Dusacq, Librairie agricole de la Maison rustique; [1850?]. vi, [107 unnumb. p. of text]. 340 [i.e.336] pl. 57cm.

NOTE. Duverger, pr. Preface by Laurillard is dated 20/XII/1849. Plates 309, 310, 324, 325 were not published. Issued also with a different title page, rec.654. The Library of the University of California, Berkeley, has the title pages for livraisons 13–15 and 22–24. Most of the drawings for the above were done by Cuvier as illustrations for a planned major work on comparative anatomy, of which the *Leçons d'anatomie comparée,* rec.710–rec.712, were a summary. Cuvier died before accomplishing his goal; cf. Laurillard, p. [i].

654. Anatomie comparée recueil de planches de myologie dessinées par Georges Cuvier ou exécutées sous ses yeux par M. Laurillard publié sous les auspices de M. le Ministre de l'instruction publique et sous la direction de MM. Laurillard . . . et Mercier . . . Paris: Imprimerie d'E. Duverger; 1850. 4 v. vi, [107] p. 340 [i.e.336] pl. 56cm.

NOTE. Plates 309, 310, 324, 325 were not published. The majority of the drawings for the above were done by Cuvier. The only apparent difference between above and rec.653 is the title page. The copy in the Smithsonian Institution Libraries is complete. The Library of the University of California, Berkeley, has an incomplete copy which is the only other copy of the issue with this title page located. That library also has an undated title page for livraisons 10–12. Cuvier planned these to be the illustrations for a major work on comparative anatomy, of which his *Leçons d'anatomie comparée* (rec.710–rec.712) were no more than a summary, cf. Laurillard, p.[i] of rec.653, but he died before undertaking the project.

655. Recherches sur les ossemens fossiles de quadrupèdes. Discours Préliminaire. [Paris; 1812?]. 120p. 26cm.

NOTE. Caption title. A separate issue of v. 1, p.1–120 of the 1812 edition of *Recherches sur les ossemens fossiles . . . ,* Paris, rec.740. A second edition was published in 1821, rec.665, and in 1822, rec.742. Later editions, most appearing separately, had the title *Discours sur les révolutions (de la surface) du globe.* The above was translated into English, rec.656–rec.661, into German, rec.662, and into Swedish, rec.664.

656. Essay on the Theory of the Earth. Translated from the French of M. Cuvier, . . . , by Robert Kerr, . . . With mineralogical notes, and an account of Cuvier's geological discoveries, By Professor Jameson. Edinburgh: William Blackwood; London: John Murray and Robert Baldwin; 1813. xiii, 265p. 2 pl. 8°.

NOTE. James Ballentyne, pr. Translation of *Discours préliminaire. . .* from the 1st edition of the *Recherches sur les ossemens fossiles . . .* rec.740. Cuvier text ends on p.193. An offset edition was published in 1971 by Gregg International Publishers, Ltd.

657. Essay on the Theory of the Earth. Translated from the French of M. Cuvier, . . . , by Robert Kerr, . . . With mineralogical notes, and an account of Cuvier's geological discoveries, by Professor Jameson. Second edition, with additions. Edinburgh: William Blackwood; London: John Murray and Robert Baldwin; 1815. xvi, 332p. 4 pl. 8°.

NOTE. Caledonian Mercury Press, Edinburgh, pr. Cuvier text is a translation of the *Discours préliminaire* . . . from *Recherches sur les ossemens fossiles* . . . Paris, 1812, rec.740, and is same as that of the 1813 translation, rec.656. Cuvier text ends on p.193.

658. Essay on the Theory of the Earth. By M. Cuvier, . . . , with mineralogical notes, and an account of Cuvier's geological discoveries, By Professor Jameson. Third edition, with additions. Edinburgh: William Blackwood; London: Baldwin, Cradock, and Joy; 1817. xxiii, 348p. 4 pl. front. 8°.

NOTE. James Ballentyne & Co., pr. Robert Kerr, translator. Cuvier text is a translation of the *Discours préliminaire* . . . from *Recherches sur les ossemens fossiles* . . . Paris, 1812, rec.740, and is same as that of the 1813 rec.656 and 1815 rec.657 translations. Cuvier text ends on p.193. Reprinted in 1978 by the Arno Press, New York.

659. Essay on the theory of the earth. By M. Cuvier, . . . with mineralogical notes, and an account of Cuvier's geological disoveries, by Professor Jameson. To which are now added, Observations on the geology of North America; illustrated by the description of various organic remains, found in that part of the world. By Samuel L. Mitchill, . . . New York: Kirk & Mercein; 1818. 431p. 8 pl. 22cm.

NOTE. W.A. Mercein, pr. Translation of the first edition, which appeared as the *Discours préliminaire* in the *Recherches sur les ossemens fossiles*. . . Paris, 1812, rec.740. Translation by Robert Kerr is same as that in rec.656, rec.657, and rec.658. Extracted in rec.661.

660. Essay on the theory of the earth. By M. Cuvier, . . . with mineralogical illustrations, by Professor Jameson. Fourth edition, with additions. Edinburgh: William Blackwood; London: T. Cadell; 1822. xxiii, 454p. front. 5 pl. 8°.

NOTE. Balfour and Clarke, pr. Robert Kerr is apparently the translator. Cuvier text is a transition between the French 1st edition, the *Discours préliminaire* in *Recherches sur les ossemens fossiles . . . ,* 1812, rec.740, and the second edition, *Discours sur la théorie de la terre,* 1821, rec.665.

661. Of changes in the structure of the earth From 'The Theory of the Earth' [and] Of the fabulous animals of the ancient writers. In: Warner, Charles Dudley, Editor. Library of the world's best literature ancient and modern. Charles Dudley Warner Edition ed. New York: The International Society; [1896]; 10: 4254–4266.

NOTE. Extracts from Kerr's translation, rec.659, of the *Discours préliminaire* in v.1 of the *Recherches sur les ossemens fossiles de quadrupèdes,* 1812, rec.740. Other editions of the above were not seen.

662. Cuvier's Ansichten von der Urwelt nach der zweiten Originalausgabe verdeutscht und mit Anmerkungen begleitet von Dr. Jakob Nöggerath, . . . [v.1]. v.2 . . . nach der dritten Originalausgabe verdeutscht und mit Anmerkungen begleitet von Dr. Jakob Nöggerath, . . . Zweiter Band. Mit zwei Steintafeln und einer Tabelle. Als Anhang: Crichton über das Clima der 6 Urwelt. Bonn: Eduard Weber; 1822–1826. [v.1],1822. [VIII], 340p. tab.; v.2,1826. xii, 222p. 2 pl. tab. 8°.

NOTE. C. F. Thormann, Bonn, pr., v.2. [V.1] is a translation of *Discours préliminaire . . .* in *Recherches sur les ossemens fossiles* . . . Paris, 1821 2nd edition, rec.742. The above was also issued separately, rec.665. V.2 contains a translation of two added chapters from the 3rd French edition, Paris, 1825, rec.666, and an update, by notes, of the Cuvier text of v.1, but neither volume contains the memoir on the ibis. The Library of Congress has the only copy of the above located in this country.

663. Rassuzhdeniia o perevorotakh na poverkhnosti zemnogo shara. Perevod s. frantsuzskogo A. E. Zhukovskogo. Redaktsiia i vstupitel'naia statiia Akad. A. A. Borisiaka. [Kommentarii A. A. Borisiaka, P. A. Novikova, G. P. Dement'eva, V. V. Struve. M.-L. Biomedgiz, I6 tipogr. Tresta "poligrafknigo"]. Moscow and Leningrad: Gosudarstvennoe

izdatel'stvo Biologicheskoi Literatury; 1937. 368p. illus., port., tab. (Klassiki biologii i meditsin').

NOTE. Added French title page, *Discours sur les révolutions de la surface du globe* 1812. Not located outside the former USSR. Not seen.

664. Idéer om Orsakerna till Jordytans närvarande form och om de Revolutioner den undergått; af Baron Cuvier, . . . Öfversättning af J. M. Stjernstolpe. Stockholm: Henrik A. Nordström; 1821. 151p.

NOTE. Swedish translation of the *Discours préliminaire,* rec.655. No copy has been located in the United States or France. Copies can be found in the libraries of Universities of Lund, Stockholm, and Uppsala and in the Eyles Collection, University of Bristol Library. Translated by J. M. Stjernstolpe. Not seen.

665. Discours sur la théorie de la terre, servant d'introduction aux recherches sur les ossemens fossiles; Par M. le B[on.] G. Cuvier, . . . Paris et Amsterdam: G. Dufour et E. d'Ocagne; 1821. clxivp. 3 pl. 4°.

NOTE. A. Belin, pr. Second edition of the *Discours préliminaire* originally published in *Recherches sur les ossemens fossiles. . . ,* Paris 1812, rec.740. Above was reissued in the second edition of the *Recherches sur les ossemens fossiles . . . ,* 1821–24, rec.742, v.2, p.239–617. Later editions have title "Discours sur les révolutions . . . " The Smithsonian Institution Libraries has the only copy of the above separate issue located in this country.

666. Discours sur les révolutions de la surface du globe, et sur les changemens qu'elles ont produits dans le règne animal; Par M. le Baron G. Cuvier, . . . Troisième édition française. Paris, Amsterdam: G. Dufours et Ed. D'Ocagne; 1825. ii, 400p. vi pl. tab. 8°.

NOTE. Fain, pr. Also published in quarto, rec.667; and also reissued a number of times. The first edition of this work was published in 1812 as the *Discours préliminaire* of the first edition in the *Recherches sur les ossemens fossiles de quadrupèdes,* rec.740. The second edition, with title *Discours sur la théorie de la terre,* was published in 1821, rec.665; and was included in the second edition of the *Recherches* . . . , rec.742. The above was also published as "Impression anastaltique," Bruxelles, Culture et Civilisation, 1969. It was translated into English, rec.683, rec.684, rec.685, and rec.745; into German, rec.686; into Italian, rec.688; and into Czech, rec.682.

667. Discours sur les révolutions de la surface du globe, et sur les changemens qu'elles ont produits dans le règne animal; Par M. le Baron G. Cuvier, . . . [3d ed.]. Paris, Amsterdam: G. Dufour et Ed. D'Ocagne; 1826. 196p. 3pl. tab. 4°.

NOTE. A. Belin, pr. Quarto printing of 3rd edition with portrait of Cuvier. Also published in octavo, rec.666. See note to rec.666 for information on other editions and for translations.

668. Discours sur les révolutions de la surface du globe, et sur les changemens qu'elles ont produits dans le règne animal; par M. le baron G. Cuvier, . . . Cinquième édition. Paris: Schubart et Heideloff; Leipzig: Ponthieu, Michelsen; 1828. vj, 400p. 6 pl. tab. 8°.

NOTE. J. Tastu, pr. Another reissue of rec.666, 3rd edition.

669. Discours sur les révolutions de la surface du globe, et sur les changemens qu'elles ont produits dans le règne animal; Par M. le Baron G. Cuvier, . . . Cinquième édition. Paris et Amsterdam: G. Dufour et Ed. D'Ocagne; 1828. vi, 400p. 6 pl. tab. 8°.

NOTE. Fain, pr. Another reissue of the 3rd edition, rec.666.

670. Discours sur les révolutions de la surface du globe, et sur les changements qu'elles ont produits dans le règne animal; Par M. le Baron G. Cuvier, . . . Sixième édition française, revue et augmentée. Paris: Heideloff; 1830. vi, 400p. 5 pl. tab. 8°.

NOTE. Casimir, pr. Reissue of rec.666. Text is that of 3rd edition.

671. Discours sur les révolutions de la surface du globe et sur les changemens qu'elles ont produits dans le règne animal Préface de Hubert Thomas Postface de Goulven Laurent Ouvrage publié avec le con-

cours du Centre national des lettres. [Paris]: Christian Bourgois; [1985]. 335p. 8°. (Collection "Épistémè" dirigée par Stéphane Deligeorges).

NOTE. Cox and Wyman, England, pr. At head of title: Georges Cuvier. Cuvier text is that of the 3rd 1825 edition, rec.666. See Note to that publication for other issues, publications, and translations.

672. Révolutions du globe précédé d'une notice par Léon Chauvin . . . Trente-neuf Gravures. Limoges: Eugène Ardant & Cie; n.d. 221p. front., illus. 28cm.

NOTE. E. Ardant et Cie, pr. At head of title: Georges Cuvier. Text is that of the 6th [i.e.4th] edition of *Discours sur les révolutions de la surface du globe,* rec.673, with minor omissions and with omission of the chapters on the zodiac and the ibis. Omissions in text are indicated. Only two copies have been located, both in private hands.

673. Discours sur les révolutions de la surface du globe, et sur les changements qu'elles ont produits dans le règne animal; Par M. le Baron G. Cuvier . . . Sixième édition française, revue et augmentée. Paris: Edmond d'Ocagne; Amsterdam: G. Dufour; 1830. ii, 408p. 6 pl. tab. 8°.

NOTE. Casimir, pr. This is actually a 4th edition consisting of the 3rd edition, rec.666, updated with insertions referring to more recent discoveries. Extracted in rec.540. Translated into German, rec.687.

674. Discours sur les révolutions de la surface du globe, et sur les changemens qu'elles ont produits dans le règne animal; par Georges Cuvier. Huitième édition. Paris: H. Cousin; Amsterdam: Veuve Legras, Imbert; 1840. 355p. 5 pl. tab. 8°.

NOTE. Edmond d'Ocagne, éditeur; Cosson, pr. Same text as 4th edition, rec.673, but omits appendix on ibis.

675. Discours sur les révolutions du globe, avec des notes et un appendice, d'après les travaux récents de MM. Humboldt, Flourens, Lyell,

Lindley, etc. . . Rédigés par le Dr. Hoefer. Paris: Firmin Didot Frères; 1850. xlvi, 276p. 4 pl. tab. 8°.

NOTE. At head of title: Cuvier. Firmin Didot, pr. Also issued: 1851, 1854, 1856, 1861, 1863, 1867, 1877, 1889. Form of publisher's name varies slightly as does that of Humboldt's on title page. Cuvier text is that of 6th edition (i.e.4th), rec.673, but does not include memoir on ibis.

676. Discours sur les révolutions de la surface du globe et sur les changements qu'elles ont produits dans le règne animal Par M. le baron Cuvier . . . Nouvelle édition conforme à la dernière publiée du vivant de l'auteur et accompagnée de notes d'après la théorie de Boitard Par F.-L. Passard. p.[iii]–iv,[1]–222; 7 pl. tab. In: Esquiros, Alphonse. L'Angleterre avant les hommes Discours sur les révolutions du globe par A. Esquiros, G. Cuvier, P.-CH. Joubert et F.-L. Passard. Paris: Delarue; [186?]. 119, 378p. illus. 8°.

NOTE. J. Claye, pr. Text by Cuvier is the same as that of the 6th (i.e. 4th) edition, 1830, rec.673, but latter lacks appendix on the Venus Hottentote. Paper cover of above has title, *L'Angleterre avant les hommes le quinzième déluge discours sur les révolutions du globe mémoire sur la Vénus Hottentote il n'y a que deux règnes dans la nature impossibilité du feu central,* etc. . . . Paris, Passard, n.d. University of California, Berkeley has the only copy of this issue located. Issued also, rec.678 and rec.679.

677. Discours sur les révolutions du globe avec des notes et un appendice d'après les travaux récents de MM. Humboldt, Flourens, Lyell, Lindley, etc. Rédigés par le Dr. Hoefer. Paris: Firmin Didot frères, fils . . . ; 1864. 356p. 4 pl. tab. 22cm.

NOTE. At head of title: Cuvier. H. Firmin Didot, pr. Cuvier text is that of 6th (i.e.4th) edition, rec.673. Also issued in 1879. The University of Georgia Library has only copy found of 1879 issue. Copy of 1879 has not been seen.

678. Discours sur les révolutions du globe, Études sur l'ibis et mémoire sur la Vénus hottentote, par G. Cuvier. Il n'y a que deux règnes dans la nature, par P.-Ch. Joubert. . . . Paris: Passard; 1864. iv, 378p. pl. 8°. (Half-title: Bibliothèque illustrée d'histoire naturelle).

NOTE. J. Claye, pr. Cuvier text on p.1–222 is same as that of rec.673, the 4th edition. The latter does not have the appendix on the Venus Hottentote. On cover: "Seule édition annotée d'après la théorie de Boitard." Published in parts. Forms part of the "Bibliothèque illustrée d'histoire naturelle." Punctuation has been added to title page for clarification. The University of Michigan has only copy located in the United States. Other issues were also published, rec.676 and rec.679.

679. Discours sur les révolutions de la surface du globe et sur les changements qu'elles ont produits dans le règne animal Par M. le baron Cuvier, . . . Nouvelle édition conforme à la dernière publiée du vivant de l'auteur et accompagnée de notes d'après la théorie de Boitard Par F.-L. Passard; Détermination des oiseaux nommés ibis par les anciens égyptiens; Extrait d'observations faites sur le cadavre d'une femme connue à Paris et à Londres sous le nom de Vénus Hottentote Par M. G. Cuvier. p.i–iv, 1–222, 7 pl. In: Esquiros, A., Cuvier, G., [et al.]. L'Angleterre avant les hommes Le quinzième déluge Discours sur les révolutions du globe Mémoire sur la Vénus Hottentote. . . . Par A. Esquiros, G. Cuvier, P. Ch. Joubert et F. L. Passard. Paris: Passard; 1864. 119, 378p. illus. 8° (Bibliothèque d'histoire naturelle).

NOTE. J. Claye, pr. Cuvier text is the same as the 6th (i.e.4th) edition of the *Discours sur les révolutions . . .*, rec.673, although rec.673 does not include the section on the Venus Hottentote. Copies of the above issue were located only at the Academy of Natural Sciences, Philadelphia, and Dartmouth College. Two other reissues were also published, rec.676 and rec.678.

680. Discours sur les révolutions du globe Par Cuvier avec des notes d'après les données les plus récentes de la science et une notice his-

torique par Paul Bory, [*pseud.* of Paul Brunet]. Paris: Berche et Tralin; 1881. xxxix, 339p. 8°.

NOTE. Burdin, Angers, pr. Located only in the library of the École Normale Supérieure and the Bibliothèque nationale in Paris. Text of the *Discours sur les révolutions* . . . is same as that of 4th edition, rec.673, but above has also 3 éloges by Cuvier and the 'Notice historique' on Cuvier. It lacks the Appendix on the Ibis, the Advertissements, and the plates and tables.

681. Discours sur les révolutions de la surface du globe. Par G. Cuvier. Erklärt von Dr. Paul Wossidlo, . . . Mit einem in den Text eingedruckten Holzschnitt. Berlin: Weidmannsche Buchhandlung; 1881. 175p. illus. 8°. 21cm.

NOTE. W. Pormetter, pr. Cuvier text, which is in French, is that of the French 6th (i.e.4th) edition, 1830, rec.673, with introductory material and commentary in German added. The British Library has only copy found.

682. Rozprawa o přewratech kůry zemnj, a o proměnách w žiwočistwu gimi způsobených, w ohledu přjrodopisném a dĕgopisném. Podlé pátého wydánj přeložil a rozmnožil Jan Swatopluk Presl, . . . Praze: Josefy Fetterlowé; 1834. viii, 318p. 4 pl. 8°.

NOTE. At head of title: Barona Gjřjho Cuviera. Czech translation by Jan Swatopluk Presl of 5th (i.e.3rd) French edition of the *Discours sur les révolutions* . . . , rec.666. Only locations found were University of Illinois Library, Urbana, Illinois; the library of the Geological Survey of Czechoslovakia (Ustrední ustav geologicky), Prague; and the Österreichische Nationalbibliothek, Vienna.

683. Essay on the theory of the earth. By Baron G. Cuvier, . . . With geological illustrations, by Professor Jameson. Fifth edition, translated from the last French edition, with numerous additions by the author and translator. Edinburgh: William Blackwood; London: T. Cadell; 1827. xxiv, 550p. 10 pl. 8°.

NOTE. P. Neill, pr. Cuvier text is a translation by Robert Kerr of the 3rd French edition, 1825, rec.666.

684. A discourse on the revolutions of the surface of the globe, and the changes thereby produced in the animal kingdom. By Baron G. Cuvier, ... Translated from the French, with illustrations and a glossary. London: Whittaker, Treacher, and Arnot; 1829. 261p. front. 3 pl. illus. 19.2 cm.

NOTE. C. and W. Reynell, pr. Translation of 3rd French edition, rec.666. This is not the same translation as rec.683, but is the same as that of rec.685.

685. Discourse on the revolutions of the surface of the globe, and the changes thereby produced in the animal kingdom. By Baron G. Cuvier, ... Translated from the French, with illustrations and a glossary. Philadelphia: Carey & Lea; 1831. iv, 252p. 6 pl. incl. front. 8°.

NOTE. Translation of 3rd French edition, rec.666. Same translation as London, 1829, rec.684.

686. Die Umwälzungen der Erdrinde in naturwissenschaftlicher und geschichtlicher Beziehung vom Baron G. Cuvier. Nach der fünften Original-Ausgabe übersetzt und mit besondern Ausführungen und Beilagen begleitet von Dr. J. Nöggerath, ... Bonn: Eduard Weber; 1830. 2 v. tab. 8° v.1, xiv, 378p.; v.2, 424p.

NOTE. C. F. Thormann, pr. Translation of *Discours sur les révolutions* . . . 5ième édition (i.e.3rd French edition), Paris, 1828, rec.666. Includes some of Professor Jameson's commentary from the *Essay on the theory of the earth*. . . , rec.683.

687. Die Erd=Umwälzungen von Georg Cuvier. Deutsch bearbeitet und mit erläuternden Bemerkungen über die neuesten Entdeckungen in der Geologie und Paläontologie versehen von C. G. Giebel, ... Mit dem Portrait Cuvier's und zwei Tabellen. Leipzig: Verlag von Ambr: Abel; 1851. xii, 274p. port. tab. 8°.

NOTE. J. S. Wassermann, pr. Translation of 6th (i.e.4th) French edition, rec.673, with last 4 chapters completely reworked (cf. preface), and without the article on the ibis.

688. Discorso del Signor Barone Cuvier su le rivoluzioni della superficie del globo Traduzione con note del sacerdote Ignazio Paradisi. Firenze: Niccolò Conti; 1828. 2v. v.1: 228p. 1 pl.; v.2: 126p. 5 pl. tab. 8°.

NOTE. Ronchi e Celli, pr. Translation of 3rd French edition, Paris, 1825, rec.666. Bound with other works, including rec.452.

689. O perevorotakh ili izmeneniiakh na poverkhnosti zemnogo shara v estestvroopisatel'nom i istoricheskom otnoshenii. Sochinenie Barona Kiuv'e. Pereved s posliedniago, dopolnennago izdaniia T. Dymchevich'. Odessa: Gor. Tip.; 1840. [8] 226, xiip.

NOTE. Not seen. Translation of the *Discours sur les révolutions de la surface du globe*. Edition of French original not identified. No location found outside the former USSR.

690. Essai sur la géographie minéralogique des environs de Paris. Par MM. G. Cuvier et Alex. Brongniart. [n.p.; 180–?]. 33p. 25cm.

NOTE. Caption title. The copy seen in the Muséum national d'histoire naturelle, Paris, was Brongniart's copy with his manuscript corrections. This is not identical with the text which appeared in 1808 in the *Annales du Museum d'histoire naturelle,* rec.241, nor with rec.691.

691. Essai sur la géographie minéralogique des environs de Paris, avec une carte géognostique, et des coupes de terrain; Par G. Cuvier, . . . ; et Alexandre Brongniart, . . . Paris: Baudouin, Imprimeur; 1811. viii, 278p. 2 pl. 1 map. 25cm.

NOTE. Originally published in *Paris. Muséum national d'histoire naturelle. Annales,* rec.241, and again, with additions, in *Académie des sciences, Mémoires,* 1810, published in 1811, rec.284, of which the present edition is a reissue with addition of a preface, cf. printed catalogue of the British Museum of Natural History Library. Text is same as that in v.1 of the *Recherches sur les ossemens fossiles . . . ,* 1812, rec.740. A later edition was published in 1822, rec.692. Cuvier's copy of the above, apparently another issue, without title page or imprint, or with title page missing, was seen at the Muséum national d'histoire naturelle, Paris. The title appears on first page of text with Brongniart's name misspelled as Brogniart. Brongniart's copy with manuscript corrrections was also seen, rec.690. Extracts from the first three chapters of the above were freely translated into German, rec.300. The above was also extracted and translated into German with translator's insertions, rec.320. The University of Illinois Library has a copy of the above in which the imprint is covered by a label, Paris, Potey. The *Bibliographie de l'Empire français, ou journal de l'imprimerie,* . . . for January 4, 1812 #1082 lists a copy, "Imprim. de Baudouin" "chez Potey," which probably refers to above.

692. Description géologique des environs de Paris, Par MM. G. Cuvier et Alex. Brongniart. Nouvelle édition, dans laquelle on a inséré la description d'un grand nombre de lieux de L'*Allemagne,* de la *Suisse,* de L'*Italie,* etc., qui présentent des terrains analogues à ceux du bassin de Paris, Par M. Alex. Brongniart; Avec 2 Cartes et 16 Planches représentant les coupes de ces terrains et beaucoup des coquilles fossiles qu'ils renferment, et une Table alphabétique de tous les lieux décrits ou seulement cités. Paris et Amsterdam: G. Dufour et E. D'Ocagne; 1822. viii, 428p. 16 pl. 2 maps. 4°.

NOTE. A. Belin, pr. A later edition of rec.691. Text is same as that in *Recherches sur les ossemens fossiles* . . . , 1821–1824, v.2, pt.2, 1822, rec.742, as well as in the 4th ed., rec.744, v.4, but above includes an additional memoir by Brongniart.

693. Description géologique des environs de Paris, Par MM. G. Cuvier et Alex. Brongniart. Troisième édition, dans laquelle on a inséré la description d'un grand nombre de lieux de l'Allemagne, de la Suisse, de l'Italie, etc., qui présentent des terrains analogues à ceux du bassin de Paris, Par M. Alex. Brongniart; avec un atlas de dix-huit planches, dont deux coloriées, et une table alphabétique de tous les lieux décrits ou seulement cités. Paris: Edmond d'Ocagne; 1835. 685p. 8° and atlas of 16 pl., and 2 maps. 4°.

NOTE. Casimir, pr. Text is same as rec.692 and as that included in *Recherches sur les ossemens fossiles* . . . , 1821–1824, v.2, pt.2, 1822, rec.742, and as in the 4th edition of the *Recherches* . . . , rec.744, although the latter has an additional memoir by Brongniart.

694. Histoire des progrès des sciences naturelles, depuis 1789 jusqu'à ce jour, par M. le Baron G. Cuvier, . . . Paris: Baudouin frères et N. Delangle; 1826–1828. 4 v. 19.9cm. [v.1],1826. 376p.; [v.2],1828. vi,419p.; [v.3], 1828. 493p.; [v.4],1828. 381p. 25cm. (Buffon, Georges Louis Leclerc, comte de. Oeuvres complètes de Buffon. Complément I–IV [29–32]).

NOTE. Jules Didot l'aîné, pr. Issued also as v.29–32. Some of above is taken from Cuvier's yearly *Analyses des travaux,* rec.567. Volume 1 was also published in 1810 with minor editorial differences, rec.733. V.1 covers the period 1789–1808, and v.2–4, 1809–1826. V.4 includes *Discours sur l'histoire naturelle . . . depuis le retour de la paix maritime,* originally published separately in 1824, rec.402. Device of medals appears on title pages. Also issued with same dates but without publication dates on title pages, and with globe and anchor as title page device and with Delangle alone as publisher of v.2–4, rec.695. Issued again in 1829 with medals and with volume numbers on title pages and Baudouin frères as publisher on all 4 volumes, rec.696. Reissued with an added volume covering 1827–1830, rec.697, and republished in 1837, rec.698, and again in 1864, rec.699.

695. Histoire des progrès des sciences naturelles, depuis 1789 jusqu'à ce jour, par M. le Baron G. Cuvier, ... Paris: Delangle frères; 1826–1828. 4 v. 22.8cm. v.[1], 1826. 376p.; v.[2],1828. vi,419p.; v.[3],1828. 493p.; v.[4],1828. 381p. (Buffon, Georges Louis Leclerc, comte de, 1707–1788. Oeuvres complètes de Buffon, . . . suivies de deux volumes sur les progrès des sciences physiques et naturelles depuis la mort de Buffon, par M. le Baron Cuvier. (Complément I–IV); v. 29–32).

NOTE. Jules Didot l'aîné, pr. V.[1] published by Baudouin frères and N. Delangle. Title page device is anchor and globe. For differences among these issues, see Note to rec.694. Reissued with a fifth volume, rec.697. V.1 of above covers 1789–1808; v.2–4 cover 1809–1826.

696. Histoire des progrès des sciences naturelles, depuis 1789 jusqu'à ce jour, par M. le Baron G. Cuvier, . . . Paris: Baudouin frères; 1829. 4v. 8° v.1, 376p.; v.2, vi, 419p.; v.3, 493p.; v.4, 381p. ([Buffon. Complément T.I–IV]).

NOTE. Jules Didot l'aîné, pr. V.1 covers 1789–1808; v.2–4 cover 1809–1826. Rec.694 and rec.695 are other issues of the above. For differences among these issues, see Note to rec.694. Reissued again with a fifth volume, rec.697. Device on title page of above is medals.

697. Histoire des progrès des sciences naturelles, depuis 1789 jusqu'à ce jour, par M. le baron Cuvier, . . . Paris: Librairie encyclopédique de Roret, Pourrat frères; 1834–1836. 5v. 8° v.1,1834, 376p.; v.2,1834, vi, 419p.; v.3,1834, 493p.; v.4,1834, 381p.; v.5,1836, 507p. (Oeuvres complètes de Buffon. Complément, v.xi–xv).

NOTE. Fain, pr. V.1–4 are a reissue of rec.694 and cover the period 1789–1826; v.5 covers 1827–1830. "1827–1831" appears on title page of v.5.

698. Histoire des progrès des sciences naturelles, depuis 1789 jusqu'à ce jour, par M. le baron G. Cuvier, . . . Bruxelles: Société belge de librairie, Hauman, Cattoir; 1837–1838. 2v. 8° v.1, 1837, 484p.; v.2, 1838, 516p.

NOTE. H. Remy, pr. Text is the same as that of 1826–1828 edition, rec.694, but has additional coverage for the period 1827–1830, found also in v.5 of rec.697. Above omits the *Discours sur l'histoire naturelle* . . . from v.4 of rec.694. Not compared with rec.697 for the period 1827–1830. Reissued in 1864, rec.699. For further details of republications, see Note to rec.694.

699. Histoire des progrès des sciences naturelles depuis 1789 jusqu'à ce jour par M. Le Baron G. Cuvier . . . Paris: Librairie internationale; Bruxelles, Livourne et Leipzig: A. Lacroix; 1864. 2v. 8° v.1, 484p.; v.2, 516p.

NOTE. Another issue of the Brussels 1837–1838 edition, rec.698. See Note for that issue for further bibliographical information.

700. Geschichte der Fortschritte in den Naturwissenschaften seit 1789 bis auf den heutigen Tag, vom G. Cuvier, ... Aus dem Französischen von F. A. Wiese. Leipzig: Baumgartner's Buchhandlung; 1828–1829. 4 v. 20.2cm. v.1,1828. x, 322p.; v.2,1829. x, 345p.; v.3,1829. x, 414p.; v.4,1829. viii, 356p.

NOTE. Not seen. Georg Maret, pr., v.1, 2. Apparently a translation of rec.694.

701. Cours de l'histoire des sciences naturelles, professé au collège de France. Par M. le baron Cuvier; rédigé par M. Magdeleine de Saint-Agy. Paris: Béchet, fils and Béchet, aîné; 1830–1833; 3 Pts.

NOTE. This publication of Cuvier's lectures on natural history completed by Magdeleine de Saint-Agy was not seen nor located. Evidence for its existence appears in the *Bibliographie de la France Journal général de l'imprimerie et de la librairie*, v.19–22, and the title pages of the above have been added to another printing, rec.702. There are various printers for the livraisons: Huzard-Courcier, Courcier, Alfred Courcier, veuve Poussin, Baudouin. Title and publisher vary slightly, but the family Béchet appears as publishers wherever a publisher is given.

702. Histoire des sciences naturelles, depuis leur origine jusqu'à nos jours, chez tous les peuples connus, professée au Collége [!] de France, par Georges Cuvier, complétée, rédigée, annotée et publiée par M. Magdeleine de Saint Agy. Paris: Fortin, Masson; 1841–1845. 5v. 8° v.1,1841. Première partie,—Comprenant les siècles antérieurs au 16e de notre ère. iii, 441p.; v.2,1841. Deuxième partie,—Comprenant les 16e et 17e siècles. 558p.; v.3,1841. Troisième partie,—Comprenant la première moitié du 18e siècle. 338p.; v.4, 1843. Troisième partie,—Contenant la deuxième moitié du 18e siècle. 351p.; v.5,1845. Troisième partie,—Contenant la fin de la deuxième moitié du 18e siècle et une partie du 19e. Tome cinquième complémentaire. 440p.

NOTE. Based on lectures delivered by Cuvier at the Collège de France. Terzuolo, pr., v.1–3; L. Martinet, pr., v.4; Bourgogne et Martinet, pr., v.5. v.4–5, have subtitle, "Commencée au Collége [!] de France par Georges Cuvier, complétée par M. Magdeleine de Saint Agy." V.1–3 in the copy of the above seen in the Bibliothèque Nationale also have cancelled title pages with publisher, Béchet, and publication dates of 1831 for v.1 and v. 2; and 1832 for v.3. See also rec.701. Cuvier sent a letter to the editor of the *Moniteur* requesting publication of the following: "La manière dont on vient d'annoncer dans tous les journaux l'impression de mes leçons au Collège de France me fait un devoir de prévenir le public que je ne prends aucune part ni à la rédaction, ni à la révision de ces sténographies, et que je ne puis nullement répondre à des erreurs qu'un pareil mode de publication doit rendre inévitable . . . ," *Le Moniteur,* April 22, 1830, no.112, p.442. In v.5 of the above, M. Magdeleine de St. Agy states: " . . . le premier volume de cet ouvrage, une partie du troisième, le quatrième et le cinquième, sont mon travail personnel. Des sténographes ont recueilli le deuxième et une partie du troisième; mais ils les ont recueillis si inexactement, j'ai été obligé de les remanier si profondément, et il y est resté si peu de phrases prononcées par le célèbre professeur . . . que la loyauté m'oblige d'en prendre aussi toute la responsabilité . . . " cf. v.5, p.434. Louis Roule, however, in his article *Cuvier, historien scientifique;* Paris, Muséum national d'histoire naturelle, Archives VI, v.9, p.78 states "Comprenant cinq volumes . . . , les deux premiers reflètent vraiment le cours de Cuvier et le reproduisent, revu par le professeur; les trois derniers sont l'oeuvre du rédacteur, aidé de ses notes, mais livré à lui seul." Most of these lectures were published in *Le Globe* (rec.480), *Le Temps* (rec.483), and *Le National* (rec.482) shortly after their delivery by Cuvier. Those compared with the above were condensed in wording and contained some omissions, especially in *Le Temps*. *Le Globe* stopped reporting the lectures in 1830, and most issues of *Le Temps* for late 1831 and 1832 are missing from the microfilm of the Bibliothèque Nationale copy, the only copy located, and therefore a more complete comparison was not made with Magdeleine de St. Agy's reworked version. The above edition was also reproduced: "Impression anastaltique," Bruxelles, Culture et Civilisation, 1969. Translated into Polish, rec.704. Selections were also translated into Polish, rec.560.

703. Histoire des sciences naturelles pendant la deuxième moitié du XVIIIe siècle et une partie du XIXe chez tous les peuples connus commencée au Collège de France par Georges Cuvier complétée par M. Magdeleine de Saint Agy. Paris: Fortin, Masson; 1843. 2v. 8° v.1, 351p.; v.2, 440p.

NOTE. Pommeret et Moreau, pr. The above is another issue of v.4 and 5 of *Histoire des sciences naturelles, depuis leur origine jusqu'à nos jours,* rec.702, with new title pages.

704. Historya nauk przyrodzonych podług ustnego wykładu Jerzego Kiuwiera (Cuvier) ułożona i uzupełniona przez p. Madelen de St.-Agy; na język polski przełożyli i dodatkami do piśmien-nictwa polskiego odnoszącemi się wzbogacili Gustaw Belke i Alexander Kremer. Wilno: Josefa Zawadskiego; 1853–1855. 5v. 8° v.1, 1853, xii, 275p.; v.2, 1854, viii, 514p.; v.3, 1854, viii, 216p.; v.4, 1854, viii, 324p.; v.5, 1855, xvi, 288p.

NOTE. Translation of rec.702, Paris, 1841–1845. Not located outside of Poland and not seen. The Libraries of the Polish Academy of Sciences, Warsaw; the Polish Academy of Sciences, Institute of zoology, Warsaw; Polskiej Akademii nauk, Krakow; the Biblioteka Narodiwa, Warsaw; and the Biblioteka Uniwersytecka, Warsaw have copies. Selections from rec.702 were also translated into Polish, rec.560.

705. Histoire naturelle des poissons, ouvrage Contenant plus de cinq mille espèces de ces animaux, décrites d'après nature et distribuées conformément à leurs rapports d'organisation, avec des observations sur leur anatomie et des recherches critiques sur leur nomenclature ancienne et moderne; Par M. le Baron Cuvier, . . . et Par M. Valenciennes, . . .

Prospectus rédigé par M. le Baron Cuvier. [Paris: Imprimerie de Mallet-Bachelier]; 1827. 24p. 22cm.

NOTE. Cover title. Issued again in 1862. Both issues have date on cover title. Publisher appears only on p.24 of 1827 printing and on cover title and p. 24 of 1862 issue. Republished with 28p., Paris and Strasbourg, F. G. Levrault [1827?] 8° with caption title and imprint at head of title and by same publisher in 4° with 4 unnumbered pages [1827?]. Published also again: Paris, P. Bertrand, [1842?], 20 pages, 8° caption title with imprint above title. Harvard University Library has only copy of the 1827 printing located in this country and the Bibliothèque nationale has the only copy of the 1862 issue located. The Levrault [1827?] 8° printing was located only in Academy of Natural Sciences, Philadelphia, and the 4° printing of the same publisher was found only in the Library of the Institut de France, Paris. The Museum of Comparative Zoology Library, Harvard University, has the only copy of the [1842?] printing located. It is possible that additional copies of these printings were bound in volumes of the complete work. The Prospectus was also printed in the *Annales du Muséum d'histoire naturelle,* 1827, with an omission, rec.424. The latter was translated into German, rec.453.

706. Histoire naturelle des poissons, par M. le B$^{on.}$ Cuvier, . . . et par M. Valenciennes, . . . Paris et Strasbourg: F. G. Levrault; Bruxelles: Librairie parisienne; 1828–1849. 22 v. 621 pl. 21cm. v.1,1828. xvj, 573p. 8pl.; v.2,1828. xxj, 490p. 32pl.; v.3,1829. xxviij, 500p. 30pl.; v.4,1829. xxvj, 518p. 27pl.; v.5,1830. xxviij, 499p. 41pl.; v.6,1830. xxiv, 559p. 34pl.; v.7,1831. xxix, 531p. 39pl.; v.8,1831. xix, 509p. 34pl.; v.9,1833. xxix, 512p. 32pl.; v.10,1835. xxiv, 482p. 23pl.; v.11,1836. xx, 506p. 26pl.; v.12,1837. xxiv, 507p. 19pl.; v.13,1839. xix, 505p. 20pl.; v.14,1839. xxij, 464p. 32pl.; v.15,1840. xxxj, 540p. 32pl.; v.16,1842. xx, 472p. 32pl.; v.17,1844. xxiij, 497p. 33pl.; v.18,1846. xix, 505p. 31pl.; v.19,1846. xix, 544p. 37pl.; v.20,1847. xviij, 472p. 16pl.; v.21,1848. xiv, 536p. 27pl; v.22,1849. xx, 532p., 91 unnumb. p. 16pl.

NOTE. Publisher, v.13–14, Pitois-Levrault; v.15, Pitois, Ve. Levrault; v.16–22, P. Bertrand, Ve. Levrault. Printer, v.1–13, F. G. Levrault, Strasbourg; v.14–22, Ve. Levrault, Strasbourg. Plate count includes double and folding plates and was made by plate number. It represents copies in the Muséum national d'histoire naturelle, Paris, and the British Library, London. The last numbered plate is 650. Published also in 4°, rec.707, which differs only in setting of text. Translated into Italian, rec.708. Cuvier died in 1832 and his contributions stopped with v.9. The table of contents indicates the sections written by each author. See especially *The author-ship of names proposed in Cuvier and Valenciennes* . . . by Reeve M. Bailey in *Copeia,* August 31, 1951; (3): 249–251 which gives authorship and dates of publication and discusses some issues of authorship. See also Sherborn, C. D. *Dates of publication*. . . in *Annals and magazine of natural history,* IX, v.15: 600, 1925, used also in Bailey above. Théodore Monod, *Mélanges ichthyologiques* in Institut français de l'Afrique Noire, Mémoires, v.68 (1965): 9–45 gives more bibliographical details including page numbers for the individual sections.

707. Histoire naturelle des poissons, par M. le B[on.] Cuvier, . . . et par M. Valenciennes, . . . Paris, Strasbourg: F. G. Levrault; Bruxelles: Librairie parisienne; 1828–1849. 22 v. fronts. 621 pl. 25.5cm. v.1,1828. xiv, 422p. 8pl.; v.2,1828. xvij, 371p. 22pl.; v.3,1829. xxij, 368p. 30pl.; v.4,1829. xx, 379p. 27pl.; v.5,1830. xx, 374p. 41pl.; v.6, 1830. xviij, 420p. 34pl.; v.7,1831. xxij, 399p. 39pl.; v.8, 1831. xv, 375p. 34pl.; v.9,1833. xxiv, 379p. 32pl; v.10, 1835. xix, 358p. 23pl.; v.11,1836. xv, 373p. 26pl.; v.12, 1837. xx, 377p. 19pl.; v.13,1839. xvij, 370p. 20pl.; v.14, 1839. xx, 344p. 32pl.; v.15,1840. xxiv, 397p. 32pl.; v.16, 1842. xviij, 363p. 32pl.; v.17,1844. xx, 370p. 33pl.; v.18, 1846. xviij, 375p. 31pl.; v.19,1846. xv, 391p. 37pl.; v.20, 1847. xiv, 346p. 16pl.; v.21,1848. xiij, 391p. 27pl.; v.22, 1849. xvj, 81, 395p. 16pl.

NOTE. Publisher, v.13–14, Pitois-Levrault, Paris; v.15, Pitois, Paris; Ve. Levrault, Strasbourg; v.16–22, P. Bertrand, Paris; Ve. Levrault, Strasbourg. Printer, v.1–13, F. G. Levrault, Strasbourg; v.14–17, Ve. Berger-Levrault, Strasbourg; v.18–22, Ve. Levrault, Strasbourg. Published also in 8° printing, rec.706, and translated into Italian, rec.708. Plates were counted in various libraries and total 621 pl. Double plates were recorded as one. Last numbered plate is 650. Reimpression, Amsterdam, A. Asher, 1969. For information on dating and contributions of individual authors, see Note to rec.706.

708. Storia naturale dei pesci del signor Barone Cuvier . . . e del signor Valenciennes . . . Riprodotta in italiano da Giuseppe de Ceresa . . . Milano: L'Editore Antonio Locatelli; v.2, 1830. 328+[?] p. 124 pl. (Locatelli, Antonio, ed. Il Regno animale tratto dalle migliori opere, cominciando dalle tre seguenti . . . ; v. 2).

NOTE. F. Sonzogno, pr. Not seen, but apparently a translation of rec.706. The only copy located is in the Universitätsbibliothek, Wien, and is incomplete, ending on p.328 in the middle of a line. The publication date on the title page for the 3[?] vol. set is 1843. The other volumes in this set are not by Cuvier.

709. Historia natural de las aves sacada de Buffon, Cuvier, etc. Paris and Mexico: A. Bouret é hijo; 1876. 175p. (Biblioteca de la Juventud).

NOTE. Text is not attributed to individual authors, and therefore Cuvier contribution is not identifiable.

710. Leçons d'anatomie comparée de G. Cuvier, . . . Recueillies et publiées sous ses yeux, par C. Duméril, . . . Paris: Baudouin, Imprimeur; An VIII [1800]. 2v. tab. 21cm. v.1, Contenant les organes du mouvement. xxxj, 521p. 7 fold. tab. v.2, Contenant les organes des sensations. xvj, 697p.

NOTE. Presented to the Académie des sciences, March 31, 1800. This set was continued by the publication in 1805 of volumes 3 to 5, with Genets as publisher, rec.711. All 5 volumes were also reissued in 1805 by Baudouin, rec.712. Volume 5 of both of these sets contains the plates for the first two volumes above. These were based on Cuvier's course on comparative anatomy, delivered at the Muséum national d'histoire naturelle, reworked with Cuvier from notes collected and edited by A. M. C. Duméril and later G. L. Duvernoy. A summary of the above two lectures appeared in 1800, rec.69. The above was translated into English, rec.716, and abridged and translated into English, rec.715 and rec.717. Translated into German, rec.718. The above includes a letter to Jean Claude Mertrud in v.1: i–xxij, also listed in rec.894, which gives information on the various printings of the letter. See also Note to rec.712.

711. Leçons d'anatomie comparée, de G. Cuvier, . . . Recueillies et publiées sous ses yeux par G.-L. Duvernoy, . . . Paris: Genets, jeune; An XIV–1805. 3 v. (v.3–5). 52 pl. 8° v.3, Contenant la première partie des organes de la digestion. xxviij, 558p. v.4, Contenant la suite des organes de la digestion et ceux de la circulation, de la respiration et de la voix. xij, 539p. v.5, Contenant Les organes de la génération et ceux des sécrétions, excrémentielles ou des excrétions. vij, 368p. 52 pl.

NOTE. Printer not indicated. Another issue of volumes 3–5 of rec.712. As specified by the publisher, these are the continuing volumes begun with first two volumes published by Baudouin in 1800 (rec.710). Plates of all 5 volumes are published in v.5 of the combined set. See Notes to rec.710 and rec.712 for further information on above. Also included in v.3, p.ix–xxviij of the above is a letter to Étienne Lacépède. See Note to rec.892 for further publication of the letter in various printings of the above.

712. Leçons d'anatomie comparée de G. Cuvier, . . . Recueillies et publiées sous ses yeux par C. Duméril, . . . [t.3–5 par G. L. Duvernoy]. Paris: Baudouin, Imprimeur; An XIV. 1805. 5 v. 52 pl. tab. 25.2cm. v.1, Contenant les organes du mouvement. xxxj, 521p. tab. v.2, Contenant les organes des sensations. xvj, 697p. v.3, Contenant la première partie des organes de la digestion. xxviij, 558p. tab. v.4, Contenant la suite des organes de la digestion et ceux de la circulation, de la respiration et de la voix. xij, 539p. tab. v.5, Contenant les organes de la génération et ceux des sécrétions excrémentielles ou des excrétions. vij, 368, 52 pl.

NOTE. Crochard, and Fantin, Libraires, also appear on the title page. Published from notes collected and edited by Duméril and later by Duvernoy of Cuvier's popular series of lectures on comparative anatomy. Volumes 1 and 2 were also issued in 1800, rec.710; and volumes 3–5 were issued by Genets in 1805, rec.711. Published also as "Impression anastaltique," Bruxelles, Culture et Civilisation, in 1969. An augmented edition of the above was published, rec.713 and rec.714. The above was translated into English, rec.716, excerpted in English, rec.717, and abridged, rec.715. Translated also into German, rec.719, rec.720, and rec.721. A letter to Jean Claude Mertrud is published in the above, v.1: i–xxij, and listed also as rec.894 with information in the Note on the various republications of the letter. See also the Note to rec.892 for information on a letter to Étienne Lacépède which appeared in above and other republications.

713. Leçons d'anatomie comparée de Georges Cuvier, recueillies et publiées par M. Duméril. [v.4–6 par M. G. L. Duvernoy]. Seconde édition, corrigée et augmentée. Paris: Crochard; 1835–1846. 8 v. in 9. tab. 8°. v.1, 1835. Contenant les généralités, et les organes du mouvement des animaux vertébrés. xxxiij, 587p. v.2, 1837. Contenant les organes du mouvement des animaux sans vertèbres, et l'ostéologie de la tête. viii, 726p. v.3, 1845. Contenant le système nerveux et les organes des sens. iv, 760p. v.4, 1835. Pt.1, Contenant les organes de mastication, d'insalivation et de deglutition des animaux vertébrés. xxij[i.e.xxvij] 628p. Pt.2, Suite de l'appareil de chylification des animaux vertébrés. 691p. v.5, 1837. Contenant les organes d'alimentation des mollusques, des animaux articulés et des zoophytes. viii, 503p. v.6, 1839. Contenant la description du fluide nourricier, de ses réservoirs et des organes qui le mettent en mouvement, dans les quatre types du règne animal. vii, 559p. v.7, 1840. Contenant la description des organes d'élaboration et de dépuration du fluide nourricier, par la respiration et la sécrétion urinaire. vii, 656p. v.8, 1846. Contenant les organes de la génération et des sécrétions, avec une leçon complémentaire des organes de relations. xii, 848p.

NOTE. Vols.3, 7, 8 published by Fortin, Masson. Printer for v.1, 4 is Hippolyte Tilliard; for v.2, 5, Felix Locquin; for v.3, 8, Bourgogne et Martinet; for v.6, 7, Terzuolo. The following appears on title pages; v.1, "revu par M. G. Cuvier"; v.2–3, "revu par MM. F. G.

Cuvier et Laurillard"; v.4–6, "revu par G. L. Duvernoy"; v.7, "revu et entièrement refondu par G.-L Duvernoy"; v.8, "par Georges Cuvier et G.-L. Duvernoy." Volumes 3 and 8 were published also in Leipzig. The above is an augmented edition of rec.712. It was also published with the same text in Brussels, rec.714. Partially translated into German, rec.722. See Notes to rec.710 and rec.712 for more information on the publication of this work. Includes letters to Jean Claude Mertrud (in v.1: xiij–xxxiij) also listed separately as rec.894; and to Étienne Lacépède (in v.4: xiij–xxij) also listed as rec.892.

714. Leçons d'anatomie comparée de Georges Cuvier, recueillies et publiées Par M. Duméril. Troisième édition. Revue, corrigée et considérablement augmentée. Bruxelles: H. Dumont; et Londres: Dulau; 1836–1840. 3 v. 29cm. v.1,1836. xv, 655p.; v.2,1838. xiv, 591p. tab.; v.3,1840. 643p.

NOTE. J. Stienon, pr., v.1. V.2–3 were also published by the Société typographique belge, Adolphe Wahlen. Text is same as that of rec.713. Incomplete copies of v.1 and v.3 of above were also seen—2 copies of v.1 ending on p.416, and one copy of v.3 ending on p.368. Record 892 also lists the letter published in above, v.2: ix–xiv, from Cuvier to Étienne Lacépède; and rec.894 lists the letter to Jean Claude Mertrud, in v.1: ix–xv of above. The Notes to these two records contain bibliographic information for other publication of these letters in the various printings of the above.

715. An introduction to the study of the animal economy; translated from the French of Cuvier, by John Allen . . . Edinburgh: Ross and Blackwood; London: Longman and Rees; 1801. 79p. 8°.

NOTE. C. Stewart, pr. Abridged translation of the "Première leçon" of rec.710 intended for the use of students.

716. Lectures on comparative anatomy. Translated from the French of G. Cuvier, . . . By William Ross; under the inspection of James Macartney, . . . London: T. N. Longman and O. Rees; 1802. 2 v. 8° v.1,

On the organs of motion. xl, 542p. 7 fold. tab.; v.2, On the organs of sensation. xx, 710p.

NOTE. Printed at the Oriental Press, by Wilson and Co. Translation of rec.710: "I have taken the liberty of correcting some errors in the original . . . ; but in general the strictest regard has been paid to the Author's mode of description and expression, . . . "; cf. preface by Macartney, p.vii. Some libraries have catalogued the tables in this separately (or they were separately published) under the title *Synoptic tables of zoology, exhibiting a classification of animals, according to their internal structure.*

717. [Excerpts from *Leçons d'anatomie comparée* translated into English]. In: Blumenbach, Johann Friedrich. A short system of comparative anatomy, translated from the German of J. F. Blumenbach, . . . by William Lawrence, . . . with numerous additional notes, and an introductory view of the classification of animals, by the translator. London: Longman, Hurst, Rees, and Orme; 1807. xxxiii, 484p.

NOTE. Strahan and Preston, pr. Translated excerpts from Cuvier, rec.712, appear for the most part at ends of chapters. "I have taken the liberty of adding notes to such parts as appeared defective either from omission, or too great conciseness; and I have placed these at the end of each chapter. Many of these are derived from Cuvier's work, which I acknowledge in this general manner, to save the trouble of numerous references." W. Lawrence on p.ix.

718. G. Cuvier . . . Vorlesungen über vergleichende Anatomie gesammelt und unter seinen Augen herausgegeben von C. Duméril, . . . Aus dem Französischen übersetzt und mit Zusätzen versehen von Gotthelf Fischer, . . . Braunschweig: Friedrich Vieweg; 1801–1802. 2 v. 8° v.1, Enthält die Organe der Bewegung. xxxvi, 620[i.e.623]p.; v.2, Enthält die Organe der Empfindungen. xviii, 749p.

NOTE. Translation of rec.710.

719. Vorlesungen über vergleichende Anatomie von G. Cuvier . . . Gesammelt und unter seinen Augen herausgegeben von C. Düméril [v.3,

4 von G. L. Duvernoy . . .]. Leipzig: Paul Gotthelf Kummer; 1809–1810. 4 v. 8° v.1, 1809, Welcher die Organe der Bewegung enthält. Uebersetzt und mit Anmerkungen und Zusätzen vermehrt von L. H. Froriep . . . und von I. F. Meckel . . . Mit sechs Kupfertafeln. xxxii, 512p. tab. 6 pl.; v.2, 1809, Welcher die Organe der Empfindungen enthält. Uebersetzt und mit Anmerkungen und Zusätzen vermehrt von I. F. Meckel . . . Mit sieben Kupfertafeln. xvi, 696p. 7 pl.; v.3, 1810, Welcher die Organe der Verdauung enthält. Uebersetzt und mit Anmerkungen und Zusätzen vermehrt von I. F. Meckel . . . Mit acht Kupfertafeln. xxxii, 743p. 8 pl. v.4, 1810, (Vierter und letzter Theil) welcher die Organe des Kreislaufs, des Athmens, der Stimme und die Generation enthält. Uebersetzt und mit Anmerkungen und Zusätzen vermehrt von I. F. Meckel . . . Mit vier Kupfertafeln. xx, 734p. 4 pl.

NOTE. J. Ch. Hendel, pr., v.1, 3, 4. V.1, tr. and ed. by L. H. Froriep and I. F. Meckel; v.2–4, tr. and ed. by I. F. Meckel. Translation of rec.712. Index to above was published in 1824, rec.720. V.4, p. 247–250 of above were republished in 1817 with comments presumably by Oken, rec.331.

720. Alphabethisches und Systematisches Register zu Cuvier's Vorlesungen über vergleichende Anatomie von F. O. Lietzau. Leipzig: Paul Gotthelf Kummer; 1824. vi, 141 p. tab. 8°.

NOTE. Index to *Vorlesungen über vergleichende Anatomie. . .* Leipzig, Kummer, 1809–1810, rec.719.

721. Vorlesungen über vergleich. Anatomie: aus d. Französ v. L. F. Froriep u. J. F. Meckel. Leipzig und Riga: Kummer; 1809–1811. 4 v. 8°.

NOTE. Not seen, nor located. Reference comes from *Gesamtverzeichnis des deutschsprachigen Schrifttums (GV),* 1700–1910, v.26, p.322. The above may be another printing of rec.719 with a change of date and the addition of Riga on the title page. W. Heinsius, *Allgemeines Bücher-Lexicon,* 1700–1810, p. 634 has another listing: Riga, Hartmann, 1810, which was also not located nor verified.

722. Vorlesungen über vergleichende Anatomie von Georg Cuvier. Gesammelt und herausgegeben von Duméril und Duvernoy. Zweite verbesserte und vermehrte Ausgabe. Erster Theil. Neu bearbeitet: die 1ste Abtheilung von G. Cuvier, die 2te Abtheilung von F. G. Cuvier und Laurillard. Uebersetzt von Dr. G. Duvernoy. Stuttgart: Hoffmann'sche Verlags-Buchandlung; 1839. xiv, 653p. 22.8cm.

NOTE. No more published? Translation of rec.713 through 8th lesson, article 7. Additions made by editors in 2te Abteilung are indicated by brackets.

723. Mémoires pour servir à celui qui fera mon éloge, écrits au crayon dans ma voiture pendant mes courses en 1822 et 1823: cependant les dates sont prises sur des pièces authentiques. In: Flourens, Pierre [i.e. Marie Jean Pierre]. Recueil des éloges historiques lus dans les séances publiques de l'Académie des sciences . . . Première série. Paris: Garnier frères; 1856: 169–193.

NOTE. F. Claye, pr. Incomplete autobiography of Georges Cuvier through 1817.

724. Mémoires pour servir à l'histoire et à l'anatomie des mollusques, Par M. le Chevalier Cuvier, . . . Avec 35 planches en taille-douce. Paris: Deterville; 1817. viii, various pagings totaling 480p. 35 pl. 4°.

NOTE. Leblanc, pr. Baudouin also appears on last page of Article 22 as printer. Some copies include an additional article, "Anatifes (Lepas)," 17p. and 1 pl., which is not included in the index. Most of the articles were originally published in the *Muséum national d'histoire naturelle, Annales,* and appear chronologically in the journal section of this bibliography. Some minor variations and misprints appear in the *Annales*. Article XVIII, for which no publication as an article was found, was translated into German, rec.360. Cuvier presented a copy of the above to the Académie des sciences at the session of November 25, 1816. Reproduced in 1969, Bruxelles, Culture et Civilisation, "Impression anastaltique."

725. Ménagerie du Muséum National d'Histoire Naturelle, ou les animaux vivans, Peints d'après nature, sur vélin, par le citoyen Maréchal, . . . Et gravés au Jardin des plantes, avec l'agrément de l'administration, par le citoyen Miger, . . . , Avec une note descriptive et historique pour chaque animal, par un Naturaliste. Paris: Miger, Dentu, Grandcher; An IX. [1800]. [20]p. 4 pl. fol.

NOTE. Baudouin, pr. Not seen by compiler. First edition, consisting only of the 1st livraison. According to the Prospectus, the text was entirely the work of Cuvier. Later editions and translations (rec.806, rec.807, rec.808, rec.809, rec.810) include text by Lacépède, and the work appears under his name in other bibliographies and catalogues. Only 2 copies of the above have been located; one in the Artis-Bibliotheek, Amsterdam and the other in the Bibliothèque Nationale, Paris. For above information, further discussion, more complete bibliographical information, and dating of this and later editions including a copy of the Prospectus, see Pieters, F.F.J.M. "La Ménagerie du Muséum national d'histoire naturelle, par Lacépède, Cuvier et Geoffroy. I. Historique de l'impression et description bibliographique des éditions françaises in-folio," *Bijdragen tot de Dierkunde,* 51(2): 219–249, 1981.

726. [Nature]. In: Lee, Sarah (Wallis) Bowdich "Mrs. R. Lee." Memoirs of Baron Cuvier. London: Longman, Rees, Orme, Brown, Green, & Longman; 1833: 139–151.

NOTE. Translation, with indicated omissions, of rec.861, Cuvier's article on nature in the *Dictionnaire des sciences naturelles.* Published also in the New York edition of above; also published in 1833, p.77–83.

727. Rapport sur les établissemens d'instruction publique en Hollande, et sur les moyens de les réunir à l'Université impériale; Fait en exécution de l'article 50 du décret impériale du 18 octobre 1811; Par M. Cuvier . . . , et par M. Noel, . . . [Paris: Fain; 181?]. 198p. 20cm.

NOTE. Fain, pr. At head of title: Université impériale. English translation, rec.729, has omissions. Not compared with rec.728.

728. Première partie du rapport sur les établissemens d'instruction publique, en Hollande, par M. Cuvier, . . . , et M. Noël, . . . Écoles primaires. (Deuxième edition). La Haye; 1818. 46p.

NOTE. Not seen. The only copy located is in the Universiteits-Bibliotheek, Amsterdam. This is possibly the original from which rec.729 was translated into English.

729. Part First of the Report on the establishments for Public Instruction in Holland, by M. Cuvier, Member of the Council; and M. Noel, . . .–1811. In: Cousin, Victor. On the state of education in Holland, as regards schools for the working classes and for the poor . . . Translated, with preliminary observations . . . by Leonard Horner, . . . London: John Murray; 1838: 257–294.

NOTE. G. Woodfall, pr. Translated extracts from rec.727. Not compared with rec.728.

730. Rapports sur les établissemens d'instruction publique des départemens au-delà des Alpes; Faits en 1809 et 1810, par une commission extraordinaire composée de MM. Cuvier, . . . ; de Coiffier, . . . ; et de Balbe, . . . Extrait du Rapport Général sur l'Académie de Gênes. [Paris: Fain; 181?]. 239p. 23.5cm.

NOTE. At head of title: Université impériale.

731. Rapport sur l'instruction publique dans les nouveaux départemens de la basse Allemagne; Fait en exécution du décret impérial du 13 décembre 1810, Par M. Cuvier, . . . , et par M. Noel, . . . [Paris: Fain; 181?]. 116p. 4°.

NOTE. At head of title: Université impériale. Excerpts translated into German, rec.563.

732. [Rapport] Discours de M. Cuvier, . . . In: Institut de France. Discours sur les progrès des sciences, lettres et arts, depuis

MDCCLXXXIX jusqu'à ce jour; ou compte rendu par l'Institut de France à S. M. l'Empereur et roi. Avec des notes sur les savans cités dans les rapports et la notice raisonnée de leurs travaux, dans lesquelles on a fait mention des ouvrages. [La Haye? Rotterdam?], Hollande: Immerzeel; 1809: p.46–80. xiv, 420p. 8°.

NOTE. With notes by J. L. Kesteloot. Also printed with the addition to above title page of 'A Paris, Chez Ant. Aug. Renuard, . . . ' Cuvier text is same as that of the *Académie des sciences, Mémoires,* 1808, rec.235, with a few short omissions. The above was expanded and published in 1810, rec.733. See Notes in rec.694 and rec.733 for more complete publishing history.

733. Rapport historique sur les progrès des sciences naturelles depuis 1789, et sur leur état actuel, Présenté à Sa Majesté l'Empereur et Roi, en son Conseil d'état, le 6 Février 1808, par la Classe des Sciences physiques et mathématiques de l'Institut, conformément à l'arrêté du Gouvernement du 13 Ventôse an X; Rédigé par M. Cuvier, . . . Imprimé par ordre de sa Majesté. Paris: L'Imprimerie impériale; 1810. xvi, 394p. 8°.

NOTE. J. J. Marcel, pr. A quarto printing was also published, rec.734. These represent a greatly expanded edition based on an original presentation published in 1808, rec.235. Text of above is same as that of rec.735 with minor editorial differences, and as that of v.1 of the *Histoire des progrès des sciences naturelles depuis 1789 . . .*, rec.694, which includes additional volumes covering later years. See Note to rec.694. The above was also reproduced in 1989 in *II Chimie et sciences de la nature,* Paris, Belin, and includes a preface by Denis Woronoff and "Présentation et notes sous la direction de Yves Laissus." The above 1810 printing was reproduced in 1969, "Impression anastaltique," Bruxelles, Culture et Civilisation and also reproduced: Amsterdam, Israel, 1970 as v.2 of series, "Spotlights on the history of science." In compliance with an order of Napoléon to prepare reports on progress since 1789 in the arts and sciences, Cuvier prepared the report on the natural sciences, complaining that it cost him what amounted to a year's work (cf. typescript of Cuvier's manuscript diary, p.25).

734. Rapport historique sur les progrès des sciences naturelles depuis 1789, et sur leur état actuel, Présenté à Sa Majesté l'Empereur et Roi, en son Conseil d'état, le 6 Février 1808, par la Classe des Sciences physiques et mathématiques de l'Institut, conformément à l'arrêté du Gouvernement du 13 Ventôse an x; Rédigé par M. Cuvier, . . . Imprimé par ordre de sa Majesté. Paris: L'Imprimerie impériale; 1810. xvi, 298p. 4°.

NOTE. J. J. Marcel, pr. An octavo edition was also published, rec.733. Above is a greatly expanded edition of the original presentation, rec.235. For further information on other editions and printings, see Note to rec.733.

735. Rapport historique sur les progrès des sciences naturelles depuis 1789, et sur leur état actuel, présenté au gouvernement, le 6 février 1808, par la Classe des sciences physiques et mathématiques de l'Institut, conformément à l'arrêté du 13 ventôse an x; Rédigé par M. Cuvier, . . . Nouvelle Édition. Paris: J. B. Baillière, Libraires; 1827. vi, 364p. 8°.

NOTE. Issued also in 1827 by L'Imprimerie Firmin Didot and in 1828 by Verdière et Lagrange, Libraires. Firmin Didot was printer of all three issues. The copy of the 1827 Didot issue seen at the Bibliothèque centrale, Muséum national d'histoire naturelle, Paris, has misprint in title ". . . leur état *naturel*, . . . " Text is same as that of 1810 edition, rec.733, with minor editorial changes and references to the Emperor omitted, and as v.1, *Histoire des progrès des sciences naturelles depuis 1789* . . . , rec.694.

736. Recherches anatomiques sur les reptiles regardés encore comme douteux par les naturalistes; faites à l'occasion de l'axolotl, rapporté par M. de Humboldt du Mexique, Par M. Cuvier. In: Humboldt, Alexander, Freiherr von. Recueil d'observations de zoologie et d'anatomie comparée; faites Dans l'océan Atlantique, dans l'intérieur du nouveau continent et dans la mer du sud, pendant les années 1799, 1800, 1801, 1802 et 1803 par A. de Humboldt et A. Bonpland. Paris: Levrault, Schoell; 1805–1809; 1: 149–196. pl.xi–xiv. 34cm. (Humboldt, Alexander and A. Bonpland. Voyage de Humboldt and Bonpland; Partie 2).

NOTE. F. G. Levrault, Strasbourg, pr. Read before the Académie des sciences on January 19 and 26, 1807. The above was published in livraison 3 of v.1, in 1807. It was republished in the complete set in 1811 [i.e.1812], rec.738. For details of publishing history of the *Recueil* . . . and the above, see C.D. Sherborn, *Annals and magazine of natural history* (7) v.3, p. 428, 1899. The above was also issued separately, rec.737. The Smithsonian Institution Libraries has the only copy of the above printing located in the United States.

737. Recherches anatomiques sur les reptiles regardés encore comme douteux par les naturalistes, faites à l'occasion de l'axolotl rapporté par M. de Humboldt du Mexique; par F.[!] G. Cuvier. Paris: L'Imprimerie de L. Haussmann; 1807. 48p. 4 pl. 35cm.

NOTE. Read to the Académie des sciences, January 19 and 26, 1807. Separate issue of rec.736.

738. Recherches anatomiques sur les reptiles Regardés encore comme douteux par les naturalistes, Faites à l'occasion de l'Axolotl Rapporté par M. de Humboldt du Mexique, Par M. Cuvier. In: Humboldt, Alexander Freiherr von. Recueil d'observation de zoologie et d'anatomie comparée: faites dans l'Océan Atlantique, dans l'intérieur du nouveau continent et dans la mer du Sud, pendant les années 1799, 1800, 1801, 1802, et 1803. Paris: F. Schoell, G.el. Dufour; 1811; 1: 93–126. pl.xi–xiv 34 cm. (Voyage de Humboldt et Bonpland; Partie 2).

NOTE. J. H. Stône, pr. Read before the Académie des sciences on January 19 and 26, 1807. The above is another and later publication of rec.736. For more complete information on the publication and dating, see Note for rec.736 and C.D. Sherborn, *Annals and magazine of natural history* (7) v.3: 428, 1899. Translated into German, rec.739.

739. Anatomische Untersuchungen über die von den Naturforschern bis jetzt für zweifelhaft gehaltenen Reptilien, angestellt bey Gelegenheit

des Axolotl, den Hr. von Humboldt aus Mexico nach Europa gebracht hat, von G. Cuvier, . . . In: Humboldt, Alexander, Freiherr von. Beobachtungen aus der Zoologie und vergleichenden Anatomie, gesammelt Auf einer Reise nach den Tropen-Ländern des neuen Kontinents, in den Jahren 1799, 1800, 1801, 1802, 1803 und 1804. Von AL von Humboldt und A. Bonpland . . . Tübingen: F. G. Cotta; Paris: Levrault, Schoell; 1806 [i.e.1808]: 160–212 4°.

NOTE. Not seen. Translation of rec.736, but not compared. The University of Rochester Library, Rochester, N.Y., has the only copy of this publication located in this country.

740. Recherches sur les ossemens fossiles de quadrupèdes, où l'on rétablit les caractères de plusieurs espèces d'animaux que les révolutions du globe paroissent avoir détruites; Par M. Cuvier, . . . Paris: Deterville; 1812. 4 v. various pagings. 27cm. v.1, Contenant le Discours préliminaire et la Géographie minéralogique des environs de Paris; 4[i.e.5]pl. map; v.2, Contenant les pachydermes des couches meubles et des terrains d'alluvion; 43pl.; v.3, Contenant les os fossiles des environs de Paris; 68pl.; v.4, Contenant les ruminans, les onguiculés et les reptiles fossiles; 38[i.e.39]pl. map.

NOTE. No printer indicated. Presented to the Académie des sciences by Cuvier on Dec. 7, 1812. Consists of separately paged memoirs originally printed in *Paris, Muséum national d'histoire naturelle, Annales*. These memoirs are listed individually and chronologically in the journal section of this bibliography. The *Discours préliminaire* in v.1 was the only original contribution to the work and is the first edition of the *Discours sur les révolutions. . .* The *Essai sur la géographie minéralogique des environs de Paris* in v.1 is a printing, with the addition of a preface, of its second edition which was published in the *Académie des sciences, Mémoires* for 1810, (1811), rec.284. The copy of the above in the Dibner Library, Smithsonian Institution Libraries, has manuscript annotations by Blainville. The above was reproduced as an "Impression anastaltique," Bruxelles, Culture et Civilisation, 1969.

741. Recherches sur les ossemens fossiles de quadrupèdes, . . . Nouvelle édition, . . . Prospectus. Paris, chez Gabriel Dufour. [Imprimerie de A. Belin?; Décembre, 1820]. 6p. 4°.

NOTE. Unsigned, but ending "Au Jardin du Roi, à Paris Décembre 1820," so presumably by Cuvier. With some minor changes, it was used as the 'Advertissement" in v.1 of the published work, rec.742 and rec.743, p. 1–5. Published also in the *Annales des sciences physiques,* Bruxelles, 1821; 8: 381–386.

742. Recherches sur les ossemens fossiles, où l'on rétablit les caractères de plusieurs animaux dont les révolutions du globe ont détruit les espèces; Par M. le $\mathrm{B}^{\mathrm{on.}}$ G. Cuvier, . . . Nouvelle édition, entièrement refondue, et considérablement augmentée. . . . Paris et Amsterdam: G. Dufour et E. D'Ocagne; 1821–1824. 5 v. pl. maps tab. 4° v.1, 1821,—Contenant le discours préliminaire, et l'histoire des éléphans, des mastodontes et des hippopotames fossiles; clxiv, 340p. 35pl.; presented to the Académie des sciences 7/30/1821 (cf. *Académie des sciences, Procès verbaux,* 7: 210). v.2, 1822, Pt.1,—Contenant l'histoire des rhinocéros, de l'elasmotherium, des chevaux, des sangliers et cochons, du daman, des tapirs et des animaux fossiles voisins des tapirs, et le résumé général de la première partie; presented to the Académie 4/1/1822 (cf. *Procès verbaux,* 7: 306). Pt.2,—Contenant la description géologique des couches des environs de Paris, parmi lesquelles se trouvent les gypses à ossemens; presented to the Académie 5/20/1822 (cf. *Procès verbaux,* 7: 327); 648p. 59pl., 2 maps. v.3,1822,—Contenant les os fossiles des environs de Paris, etc. 412p. 80pl.; presented to the Académie 7/1/1822 (cf. *Procès verbaux,* 7: 345). v.4, 1823,—Contenant les ruminans et les carnassiers fossiles; 514p. 40pl.; presented to the Académie 3/24/1823 (cf. *Procès verbaux,* 7: 465). v.5, 1823–1824,—Contenant les rongeurs, les édentes et les mammifères marins; les ossemens de reptiles et le résumé général, Pt.1, 405p. 27pl.; received by the Académie 12/22/1823 (cf. *Procès verbaux,* 7: 604). Pt.2, 547p. 33pl.; received by the Académie 12/6/1824 (cf. *Procès verbaux,* 8: 161).

NOTE. A. Belin, pr., v.1,2,4,5. Fain, pr., v.3. New edition. Contains "Additions and Corrections" to the edition of 1812, rec.740. The *"Discours préliminaire"* of this edition is the 2nd edition, rec.665. Later editions of the "Discours. . . " were also published separately under the title *Discours sur les révolutions du globe.* V.2, pt.2, p.239–617 of the above contains another issue of the *Description géologique. . .* Paris, 1822 edition, rec.692. A possible supplement to the above was published as v.XI of the *Règne animal,* rec.751.

743. Recherches sur les ossemens fossiles, où l'on rétablit les caractères de plusieurs animaux dont les révolutions du globe ont détruit les espèces; Par M. le B[on.] G. Cuvier. . . Troisième édition. Paris et Amsterdam: G. Dufour et E. d'Ocagne; 1825. 5 v. 29cm. v.1, 196, 340p. 38 pl.; v.2, iv, 648p. 59 pl., 2 maps; v.3, 412p., 88 pl.; v.4, 514p. 40 pl.; v.5, Pt.1, 405p. 27 pl.; v.5, Pt.2, 547p. 33 pl.

NOTE. A. Belin, pr. Text is a republication of the Nouvelle édition (2nd) of 1821–1824, rec.742, with the exception of the *Discours sur les révolutions* . . . in v.1 which is the 6th (i.e.4th) edition, rec.673. See rec.742 for additional information on volume titles etc.

744. Recherches sur les ossemens fossiles, où l'on rétablit les caractères de plusieurs animaux dont les révolutions du globe ont détruits les espèces; par Georges Cuvier. Quatrième Édition. Paris: Edmond d'Ocagne [etc.]; 1834–1836. 10 v. tab. 8° and atlas of 94, 106p. and 278pl. some fold., 2 maps, in 2v. 4° v.1, 1834, xxiv, 587p.; v.2, 1834, 500p.; v.3, 1834, 435p.; v.4, 1834, iv, 691p.; v.5, 1835, 687p.; v.6, 1835, 452p.; v.7, 1835, 533p.; v.8, 1836, Pt.1, 462p., Pt.2, 332p.; v.9, 1836, 501p.; v.10, 1836, 494p.

NOTE. Casimir, pr. Text is the same as that of the 1821–1824 edition, rec.742, with the exception of the *Discours sur les révolutions* . . . in v.1 which is the 6th (i.e.4th) edition, and with the additions and corrections which appeared at the end of volumes in the 1821–1824 edition incorporated into the text of the above. V.4 contains another issue of the *Description géologique* . . . Paris, 1822, rec.692.

745. Researches into fossil osteology, partially abridged and re-arranged from the French of The Baron Cuvier, . . . Part I. London: Geo. B. Whittaker; 1826. cxxviii p. 16 pl. 27.5cm.

NOTE. Davison, pr.? Cover title. No more published? Part I represents an incomplete translation of *Discours sur les révolutions, . . .*, 1825, 3rd edition, rec.666. The British Library contains the only copy of the above located.

746. Researches on fossil bones, in which are established the characters of various animals whose species have been destroyed by the revolutions of The Globe; by Baron Cuvier . . . Fourth edition, Revised and Completed by additional notes, and a supplement left by the author. In four volumes. London: G. Henderson; 1834–1835; Pts. 1–11. 484p. 23 pl., 2 fold. tab. 21.5cm.

NOTE. J. Henderson, pr.? Published in parts. No more published. Incomplete translation of 4th French edition, Paris, 1834–1836, rec.744. Plates counted were in the copy at the British Library. No copy of the above was located in the United States. Also published as micro-opaque, New York Readex microprint, 1980 (Landmarks II).

747. Le Règne animal distribué d'après son organisation, pour servir de base à l'histoire naturelle des animaux et d'introduction à l'anatomie comparée. Par M. le Cher. Cuvier, . . . Avec Figures, dessinées d'après nature . . . Paris: Deterville; 1817 [i.e.1816]. 4 v. pl. 8° v.1. Contenant l'introduction, les mammifères et les oiseaux. xxxvij, 540p.; v.2. Contenant les reptiles, les poissons, les mollusques, et les annélides. xviij, 532p.; v.3. Contenant les crustacés, les arachnides et les insectes, par M. Latreille. xxix, 653p.; v.4. Contenant les zoophytes, les tables, et les planches. viij, 255p. 15 pl.

NOTE. A. Belin, pr. Receipt of this publication was announced at the December 2, 1816 meeting of the Académie des sciences (cf. *Académie des sciences, Procès verbaux*, v.6, (1816–1819), 1915: 114). For further bibliographical information, see also C. F. Cowan, "Cuvier's *Règne animal*, First Edition" in the *Society for the Bibliography of Natural History, Journal*, 5: 219, 1969. Reproduced in 1969, "Impression anastaltique," Bruxelles, Culture et Civilisation.

Volumes 1–7 were translated into English, rec.751, and a partial translation by Bowdich into English of the mammals and birds sections was included in rec.779 and rec.780. Extracted in Polish. The French edition translated into Polish was not determined.

748. Le règne animal distribué d'après son organisation, pour servir de base à l'histoire naturelle des animaux et d'introduction à l'anatomie comparée. Par m. le baron Cuvier, . . . Avec figures dessinées d'après nature. Nouvelle édition, revue et augmentée. Paris: Deterville et Crochard; 1829–1830. 5 v. 8° v.1, 1829. xxxviij, 584p.; v.2, 1829. xv, 406p.; v.3, 1830. xvj, 504p. 20 pl.; v.4, 1829. Crustacés, arachnides et partie des insectes. Par M. Latreille. . . , xxvij, 584p.; v.5, 1829. Suite et fin des insectes. Par M. Latreille, . . . xxiv, 556p.

NOTE. Hippolyte Tilliard, pr. Second edition of rec.747. Following the French texts of the *Règne animal* listed in this bibliography are translations and translated extracts from the above edition into English, German, Hungarian, Italian, Polish, and Welsh. The above was also extracted in popular editions, rec.789–rec.794.

749. Le règne animal distribué d'après son organisation, pour servir de base à l'histoire naturelle des animaux et d'introduction à l'anatomie comparée, par M. le baron Cuvier, . . . Troisième Édition. Avec figures dessinées d'après nature. Bruxelles: Louis Hauman; 1836. 3 v. 24 pl. 8° v.1. 2, xxij, 626p. 14 pl.; v.2. 494p. 10 pl.; v.3. 518p.

NOTE. L. Schapen, Louvain, pr. Same text as second edition, rec.748.

750. Le règne animal distribué d'après son organisation, pour servir de base à l'histoire naturelle des animaux, et d'introduction à l'anatomie comparée, par Georges Cuvier. Édition accompagnée de planches gravées, representant les types de tous les genres, les caractères distinctifs des divers groupes et les modifications de structure sur lesquelles repose cette classification; par une réunion de disciples de Cuvier, MM. Audouin, Blanchard, Deshayes, Alcide d'Orbigny, Doyère, Dugès, Duvernoy, Laurillard, Milne Edwards, Roulin et Valenciennes. Paris: Fortin, Masson; [1836–1849]. 11 v. and 11 atlases. 993 pl. 4° v.[1, 1836–47].

Les mammifères. Avec un atlas, par MM. Milne Edwards, Laurillard et Roulin. xxxvj, [v]–xij, 350p., Atlas 121pl.; v.[2, 1836–43]. Les oiseaux. Avec un atlas, par M. Alcide d'Orbigny. v, 370p., Atlas 102pl.; v.[3, 1837–42]. Les reptiles. Avec un atlas, par M. Duvernoy, . . . 169p., Atlas 46pl.; v.[4, 1836–43]. Les poissons. Avec un atlas, par M. A. Valenciennes, . . . 392p., Atlas 122pl.; v.[5, 1836–45]. Les mollusques. Avec un atlas, par M. G. P. Deshayes. 266p., Atlas 152pl.; v.[6, 1836–45]. Les insectes. Avec un atlas, par MM. Audouin, Blanchard, Doyère, et Milne Edwards. Myriapodes, Thrysanoures, Parasites, Suceurs, et Coléoptères. xij, 557p., Atlas 83pl.; v.[7, 1845–49]. Les insectes. Avec un atlas, par MM. Audouin, Blanchard, Doyère, et Milne Edwards. Orthoptères, Hemiptères, Neuroptères, Hymenoptères, Lepidoptères, Rhipiptères et Diptères. 443p., Atlas 119pl.; v.[8, 1836–49]. Les arachnides. Avec un atlas, par M. Ant. Dugès, . . . et par M. Milne Edwards, . . . 106p., Atlas 31pl.; v.[9, 1836–49]. Les crustacés. Avec un atlas, par M. Milne Edwards, . . . 278p., Atlas 87pl.; v.[10, 1837–48]. Les annélides. Avec un atlas, par M. Milne Edwards, . . . 54p., Atlas 30pl.; v.[11, 1836–49]. Les zoophytes. Avec un atlas, par M. Milne Edwards . . . 160, 94p., Atlas 100pl.

NOTE. Paul Renouard, pr. The "disciples edition" published in 262 livraisons. Dates of publication and number of plates taken from C. F. Cowan, "On the Disciples Edition of Cuvier's *Règne animal," Journal of the Society for the Bibliography of Natural History*, 8: 32–64, 1976. See this article for more precise dating, authorship, and collation. Cuvier text is that of the 2nd edition, rec.748. An unnumbered page of explanation accompanies each plate.

751. The animal kingdom arranged in conformity with its organization, by the Baron Cuvier, . . . with additional descriptions of all the species hitherto named, and of many not before noticed, by Edward Griffith, . . . and others. London: Geo. B. Whittaker; 1827–1834. 15 v. fronts. pl. tab. 8°. v.1–4, 1827. The class mammalia, arranged by the Baron Cuvier with specific descriptions by Edward Griffith, Charles Hamilton Smith, . . . and Edward Pidgeon, . . . v.1. lxxx, 352p. 25pl.; v.2. 513p.

68pl.; v.3. 468p. 65pl.; v.4. 498p. 52pl.; v.5, 1827. Synopsis of the species of the class mammalia, as arranged with reference to their organization, by Cuvier and other naturalists. With specific characters, synonyma &.,&. 391p.; v.6–8, 1829. The class Aves arranged by the Baron Cuvier, with specific descriptions by Edward Griffith, . . . and Edward Pidgeon, . . . the additional species inserted in the text of Cuvier by John Edward Gray, ... v.6. 548p. 32pl.; v.7. 586p. 60pl.; v.8. 690p. 68pl.; v.9, 1831. The class Reptilia arranged by the Baron Cuvier, with specific descriptions by Edward Griffith, . . . and Edward Pidgeon; 481p. 55pl. [Followed by] A synopsis of the class Reptilia. By J. E. Gray, . . . 110p.; v.10, 1834. The class Pisces arranged by the Baron Cuvier, with supplementary additions, by Edward Griffith, . . . and Charles Hamilton Smith, . . . viii, 680p. 64pl.; [v.11], 1830. The fossil remains of the animal kingdom, by Edward Pidgeon. 544p. 49pl. tab.; v.12, 1834. The Mollusca and Radiata. Arranged by the Baron Cuvier, with supplementary additions to each order by Edward Griffith, . . . and Edward Pidgeon. viii, 601p. 61pl.; v.13, 1833. The classes Annelida, Crustacea, and Arachnida arranged by the Baron Cuvier [the last two by P. A. Latreille] with supplementary additions to each order by Edward Griffith, . . . and Edward Pidgeon. viii, 540p. 60pl.; v.14–15, 1832. The class Insecta arranged by the Baron Cuvier [by P. A. Latreille]; with supplementary additions to each order by Edward Griffith, . . . and Edward Pidgeon, and notices of new genera and species, by George Gray; v.14. viii, 570p. 53pl.; v.15. 796p. 87pl.

NOTE. William Clowes, pr., v.1–5,[11]; Shackell and Baylis, pr., v.6–8; Gilbert & Rivington, pr., v.9,10,12–16. Publisher of v.6–8,11,13–15 is Whittaker, Treacher, and of v.10, 12 is Whittaker. Issued in parts. Each volume of above has also its own title page. Some volumes of this issue also appeared as monographs, without the general title page, but with a volume title page. Title and imprint vary slightly.

Volumes 1–7 are a translation of the 1817 edition of the *Règne animal* (rec.747). Volumes 8 on, with the exception of v.[11], are translations of the 1829–1830, 2nd edition (rec.748), cf. Cowan below. Translation was probably mainly by Pidgeon, cf. Cowan. The Cuvier text is brought up to date with supplementary additions and changes. The British Museum (Natural History) Catalogue attributes v.5 to J. E. Gray, with the Ruminantia by C. H. Smith. V.11 by Pidgeon is apparently a supplement he prepared for *Recherches sur les ossemens fossiles* . . . rec.742, cf. Cowan below. For further information on dating and authorship, see Cowan, C. F. "Notes on Griffith's 'Animal Kingdom of Cuvier'," *Journal of the Society for the Bibliography of Natural History,* 5: 137–140, 1969. Above was also published as micro-opaque, New Readex Microprint, 1968.

752. [The animal kingdom described and arranged . . .], London, 1827–1834]. A classified index and synopsis of the animal kingdom arranged in conformity with its organization, by the Baron Cuvier, . . . with supplementary additions to each order, by Edward Griffith, . . . and others. London: Whittaker; 1835. 328, cxixp. 8°.

NOTE. Gilbert & Rivington, pr. Appeared as the final volume, v.[16], of the 1827–1834 edition, rec.751. Also issued in 1844, with William Stevens, pr.

753. The animal kingdom arranged in conformity with its organization, by the Baron Cuvier, . . . The Crustacea, Arachnides and Insecta, by P. A. Latreille; . . . Translated from the French, with notes and additions by H. M'Murtrie, . . . In four volumes, with plates. New York: G. & C. & H. Carvill; 1831. 4 v. 8° v.1. xxxii, 448p. 4pl.; v.2 xv, 475p. 10pl.; v.3. xx, 575p. 3pl.; v.4. xii, 545p. 3pl.

NOTE. James Kay, Jun., pr., Philadelphia. Translation by Henry McMurtrie of the 1829–1830 [2nd] French edition, rec.748. This translation was the basis for most of the English translations. Reprinted in 1833.

754. The animal kingdom, arranged in conformity with its organization. By the Baron Cuvier, . . . Translated from the French, and abridged for the use of schools, &c. by H. M'Murtrie, . . . New York: G. & C. & H. Carvill; 1832. vi, 532p. 8°.

NOTE. James Kay, Jun. & Co., pr., Philadelphia. Issued also with date 1833. Abridgement of rec.753 with more omissions than rec.755. "The whole has been sedulously, . . . and so thoroughly, expurgated that it may be placed in the hands of females, . . . "; preface, p.vi.

755. Cuvier's animal kingdom: arranged according to its organization, translated from the French, and abridged for the use of students. By H. McMurtrie, . . . With numerous illustrations on wood. London: Orr and Smith; 1834. xx, 508p. illus. 8°.

NOTE. Bradbury and Evans, pr. Abridgement of rec.753, but a fuller presentation than that of rec.754.

756. The animal kingdom, arranged according to its organization, serving as a foundation for the natural history of animals, and an introduction to comparative anatomy. By Baron Cuvier, . . . with figures designed after nature: the Crustacea, Arachnides, & Insecta, by M. Latreille, . . . Translated from the latest French Edition. With additional notes, and illustrated by nearly 500 additional plates . . . In four volumes. London: G. Henderson; 1834–1837. 4 v. ports. fronts. and atlas of 4 v. of pl. 8° v.1,1834. Mammalia–Birds. xlvii, 380p.; v.2,1834. Reptiles–Fishes. xxxv, 412p. fronts., ports.; v.3,1834. Mollusca–Annelides–Crustacea–Arachnides and Insecta. xxxv, 472p.; v.4,1836. Insecta–Zoophytes. xliv, 544p. Plates: v.1,1834. xviiip. 197pl.; v.2,1837. xxiip. 201pl.; v.3,1837. xxivp. 200pl.; v.4,1836. xxxii [i.e.xxxiii]p. 150pl.

NOTE. W. M'Dowall, pr., v.1; J. Henderson, pr., v.2–4. Added title pages, engraved, with imprint date: 1837 (v.4 varies). Atlas has engraved title pages only. Number of plates varies among copies seen. The text volumes of the Smithsonian Institution Libraries set have engraved title pages with imprint date, 1834. Minor variations in general title occur in several volumes. The count above is of the set in the Smithsonian Institution Libraries. The text of v.1 and 2 is dated 1834, but v.1 & 2, p.1–64, appeared in 1833; cf. British Museum (Natural History) Catalogue. Translation by Henry McMurtrie of the 2nd French edition, rec.748. Cuvier text is same as that of N.Y. 1831 edition, rec.753. Some of the plates were copied from Guérin-Méneville, rec.802.

The above appeared originally in 50 parts. The copy in the British Museum (Natural History) Library, 1833–1834, is a set, "in the original parts as issued. Wanting 20–25, 39 and 48"; cf. Catalogue of the British Museum (Natural History) Library. Some of the parts are not complete as in the full set above.

757. The animal kingdom of the Baron Cuvier, enlarged and adapted to the present state of zoological science. Illustrated after the original drawings of Audebert, Baraband, Cramer, d'Orbigny, Edwards, Geoffroy-St-Hilaire, Gilpin, Huet, Maréchal, Nitschmann, Oudart, Titian R. Peale, Prêtre, Redoute[!], Reinold, A. Rider, Savigny, Werner, A. Wilson, &c.&c. Edinburgh: Published for the proprietor and London: Smith, Elder, etc., etc.; [1835], 1839–1840. 2 v. fol. v.1. 198p. 95pl.; v.2. 57p. 32pl. (forms v.1 and 2, [1835–]1839–1840 of the *Edinburgh journal of natural history*).

NOTE. Edinburgh Printing Company, pr., and also publisher of v.2. Seven additional publishers in several cities are listed on title page. Text consists only of sections from the Mammals on Cuvier's Bimanes, Quadrumanes and Cheiropteres. No more published. Plates cover mammals, birds, mollusks, and insects. Translation of the 2nd French edition, rec.748, greatly expanded and with many additional species described. Translated by William Macgillivray, cf. British Library (British Museum) Catalogue.

758. The animal kingdom, arranged after its organization, forming a natural history of animals, and an introduction to comparative anatomy. By Baron Cuvier, . . . With considerable additions by W. B. Carpenter, . . . , and J. O. Westwood, . . . New edition. Illustrated with 500 engravings on wood and 36 on steel. London: Hardwicke and Bogue; [184?]. xxii, 706p. 36pl. illus. 25cm.

NOTE. William Clowes and Sons, pr. Translation of the 2nd French edition, rec.748. The above edition was reissued by a number of different publishers over a period of many years. The only copy of this issue located is in the British Museum (Natural History) Library. The text is the same as that of the 1849 edition, rec.761, translated by Henry McMurtrie, with minor additions to the preface and rearrangement of supplementary sections to follow appropriate parts of Cuvier text.

759. Cuvier's animal kingdom, Arranged according to its Organization; forming the basis for a natural history of animals, and an introduction to comparative anatomy. Mammalia, birds, and reptiles, by Edward Blyth. The fishes and radiata, by Robert Mudie. The molluscous animals, by George Johnston, . . . The articulated animals, by J. O. Westwood, . . . Illustrated by three hundred engravings on wood. London: Wm. S. Orr; 1840. vii, [2] 670p. front., illus., 26 pl. 8°.

NOTE. Baker & Darby, pr. Translation chiefly by Henry McMurtrie of the 2nd French edition, rec.748, with updates inserted in the appropriate places and indicated by brackets. Record 761 is another printing up to p.660. Most plates were missing from copies seen, although in some, guard sheets and list of plates were present. The latter was used in collation above.

760. An introductory view of the Animal Kingdom, Translated from the French of Baron Cuvier. In: Goldsmith, Oliver. A history of the earth and animated nature, by Oliver Goldsmith. . . . And copious notes embracing accounts of new discoveries in natural history; . . . and a carefully prepared index to the whole work. In two volumes. London, Edinburgh and Dublin: A Fullarton; 1847; 1: [vii]–liv.

NOTE. Fullarton and Co., pr. ". . . . a careful translation of *Baron Cuvier's Introductory View of the Animal Kingdom,* with additions from the pen of an eminent naturalist . . ," Publisher's Advertisement, p.[1]. Translation of first part of introduction to the *Règne animal* (rec.748) with many and lengthy insertions. In the title of later issues the "additions and notes" are attributed to "the late Professor MacGillivray." Issued a number of times subsequently. The Cuvier portion in some printings is numbered [ix]–lvi and is the result of a different assignment of Roman numerals to the preliminary unnumbered pages. The above issue was not compared with all of the later printings. Place of publication varies with New York added. Not all issues have been seen. Also translated into Welsh, rec.771.

761. The animal kingdom, Arranged after its Organization, forming a natural history of animals, and an introduction to comparative anatomy. By the late Baron Georges Cuvier, . . . Translated and adapted to the present state of science. The mammalia, birds, and reptiles, by Edward Blyth. The fishes and radiata, by Robert Mudie. The molluscous animals, by George Johnston, . . . The articulated animals, by J. O. Westwood, . . . A new edition, with additions by W. B. Carpenter, . . . and J. O. Westwood, . . . Illustrated by Three Hundred Engravings on Wood and Thirty=four on Steel. London: Wm. S. Orr; 1849. vii, [2], 718p. front., illus., 32 pl. 8°.

NOTE. G. Beard, pr. Added title page, engraved. Issued again in 1851 and 1854 (printer not indicated). Translation of 2nd French edition, rec.748. Translation is based on Henry McMurtrie's translation, rec.753, with only minor word changes. It is same as 1840 edition, rec.759, with the inserted updates, but has a supplemental section of updates on p. 661–708.

762. The animal kingdom, arranged after its organization; forming a natural history of animals, and an introduction to comparative anatomy. By the late Baron Cuvier, . . . Translated and adapted to the present state of science. A new edition, with additions by W. B. Carpenter . . . and J. O. Westwood . . . London and Edinburgh: A. Fullarton; 1859. xxii, 706p. illus., xliv pl. 8°.

NOTE. Fullarton & Macnab, Edinburgh, pr. Added title page, engraved. Translation of the 2nd French edition, rec.748. The British Library has a copy of this in 26 undated parts, which the British Library (British Museum) Catalogue cites as 1858–[1860]. The above is a reprint of rec.761 with interspersed supplements. Reissued by other publishers, rec.763, rec.764, and rec.765.

763. The animal kingdom, arranged after its organization; forming a natural history of animals, and an introduction to comparative anatomy.

By the late Baron Cuvier, . . . Translated and adapted to the present state of science. New edition with considerable additions by W. B. Carpenter . . . and J. O. Westwood . . . Illustrated by 500 engravings on wood and 44 on steel. London: Henry G. Bohn; 1863. xxii, 706p. front., illus., 44 pl. 8°.

NOTE. J. E. Adlard, pr. Added title page. Apparently a republication of London, 1849 edition, rec.761, with bracketed updates in the text, but with the supplement of rec.761 incorporated into the text. Translation essentially by McMurtrie, see Note to rec.761. Reprinted by Kraus, 1969.

764. The animal kingdom, arranged after its organization, forming a natural history of animals, and an introduction to comparative anatomy. By Baron Cuvier, . . . With considerable additions by W. B. Carpenter, . . . and J. O. Westwood, . . . New edition. Illustrated with 500 engravings on wood and 36 on steel. Boston: Estes and Lauriat; 1878. xxii, 706p. illus., 36 pl. 8°.

NOTE. William Clowes and sons, pr. Reissue of rec.762.

765. The animal kingdom, arranged after its organization, forming a natural history of animals, and an introduction to comparative anatomy. By Baron Cuvier, . . . With considerable additions by W. B. Carpenter, . . . and J. O. Westwood, . . . New edition, Illustrated with 500 engravings in wood and 36 on steel. London: W. H. Allen; 1886. xxii, 706p. 36 pl. 4°.

NOTE. W. H. Allen, pr. Reissue of rec.762. It was issued again by Allen in 1890. A possible 1893 printing has not been located.

766. Das Thierreich eingetheilt nach dem Bau der Thiere als Grundlage ihrer Naturgeschichte und der vergleichende Anatomie von dem Herrn Ritter von Cuvier . . . Aus dem Französischen frey übersetzt und mit vielen Zusätzen versehen von H. R. Schinz . . . Stuttgart und Tübingen: J. G. Cotta; 1821–1825. 4 v. 8° v.1, 1821. Säugethiere und Vögel. xxxviii, 894p.; v.2, 1822. Reptilien, Fische, Weichthiere,

Ringelwürmer. xvi, 835p.; v.3, 1823. Krebse, Spinnen, Insekten. xviii, 932p.; v.4, 1825. Zoophyten. xiii, 793p.

NOTE. No printer indicated. Translation of rec.747 with insertions by Schinz.

767. Das Thierreich, geordnet nach seiner Organisation. Als Grundlage der Naturgeschichte der Thiere und Einleitung in die vergleichende Anatomie. Vom Baron von Cuvier, . . . Nach der zweiten, vermehrten Ausgabe übersetzt und durch Zusätze erweitert von F. S. Voigt, . . . Leipzig: F. A. Brockhaus; 1831–1843. 6 v. 8° v.1, 1831. Die Säugethiere und Vögel enthaltend. xlviii, 975p.; v.2, 1832. Die Reptilien und Fische enthaltend. xvi, 539p.; v.3, 1834. Die Mollusken enthaltend. xviii, 621p.; v.4, 1836. Die Anneliden, Crustaceen, Arachniden und die ungeflügelten Insekten enthaltend. xiv, 516p.; v.5, 1839. Die eigentlichen Insekten enthaltend. xxvii, 686p.; v.6, 1843. Die Zoophyten enthaltend. xvi, 579p.

NOTE. F. A. Brockhaus, pr. Title indicates it is a translation of rec.748, but they have not been compared.

768. Das Thierreich geordnet nach seiner Organisation, als Grundlage der Naturgeschichte der Thiere und als Einleitung in die vergleichende Anatomie. Vom Freiherrn Georg v. Cuvier, . . . Nach der zweiten vermehrten, Ausgabe frei ins Deutsche übersetzt und durch Zusätze sowohl dem heutigen Standpunkte der Wissenschaft angepasst als auch für den Selbstunterricht eingerichtet von Aug. Vollr. Streubel. Erster Theil. Berlin: Druck und Verlag von G. Reimcr; 1846. xvii, 972p. tab. 8°.

NOTE. No more published. Translation chiefly of the Introduction in v.1 of the *Règne animal,* rec.748, interspersed with commentary by Streubel taking up a large part of the volume.

769. Az állat-ország fölosztva alkotása szerint alapul szolgálandó az állatok természetleírásához s bevezetésül az összehasonlító bonctanhoz. Cuvier Báró . . . Természet után rajzolt ábrákkal. A második átnézett és öregbített kiadás szerint fordította Vajda Péter . . . I. Kötet. Buda: A magyar királyi egyetem betüível. A. M. Academ. Költségein; 1841. lx, 644p.

NOTE. Translation of the Birds and Mammals portions of the 2nd French edition of the *Règne animal,* rec.748. These two were not compared. No more published. Translation of title: "The animal world, with sections starting with its creation, followed by a scientific description of the animals, leading into their comparative anatomies. Baron Cuvier . . . with drawings by artist of animals as in natural environment . . . the second review and older edition . . . ".

770. Il Regno animale distribuito secondo la sua organizzazione opera del Barone Cuvier, compendiata e recata in lingua italiana per servire di base alla storia naturale degli animali e d'introduzione al prodromo della *Fauna* dell'Italia superiore, compreso nei cataloghi sistematici e descrittivi delle raccolte zoologiche di G^{e}. De Cristofori e G.° Jan, . . . Parte II^{a}. I Molluschi. Parma: Stamperia Carmignani; 1832. 21p. 21cm.

NOTE. Translated extracts from the Mollusks section of the *Règne animal,* rec.748.

771. Golygiad arweiniol o'r deyrnas anifeilaidd wedi ei gyfieithu o Ffrangcaeg y Baron Cuvier, gyda nodiadau ychwanegol gan y diweddar Broffeswr Macgillivray . . . In: Goldsmith, Oliver. Hanes y Ddaear a'r creaduriaid byw, gan Oliver Goldsmith. Gyda Golygiad arweiniol o'r deyrnas anifeilaidd gan y Baron Cuvier; nodiadau helaeth o ddarganfyddiadau yn hanes natur. . . . Edinburgh a LLundain: A. Fullarton a'i gym; [1868]; 1:55–105. 26cm.

NOTE. Welsh translation of an English translation, rec.760, with its numerous and extensive insertions. It is from the first part of the introduction to the second French edition, of the *Règne animal,* rec.748. The copy in the British Library is the only copy located.

772. Tableau élémentaire de l'histoire naturelle des animaux. Par G. Cuvier . . . Paris: Baudouin, Imprimeur; an 6 [1797/1798?]. xvj, 710p. 14 pl. 8°.

NOTE. Derived from his course on natural history at the École Centrale du Panthéon. Presented to the Classe des sciences mathématiques et physiques of the Académie des Sciences, Dec. 26, 1797

(6 Nivôse an 6, cf. *Académie des sciences, Procès verbaux,* v.1: 320). A report by Lacépède and others on the above, however, was read on January 10, 1797 (21 Nivôse an 5, cf. *Académie des sciences, Procès verbaux,* v.1: 160–161). The above was translated into Danish, rec.773; into German, rec.774; Portuguese, rec.775; and Spanish, rec.776. Copy in Institut National d'Histoire Naturelle, Brussels, has imprint on slip cancel, 'Chez Theophile Barrois, le jeune'. Published also as "Impression anastaltique," Bruxelles, Culture et Civilisation, 1969.

773. Begyndelsesgrunde i dyrhistorien af Georg Cuvier, . . . Af det franske ved Gregers Wad, . . . Kiöbenhavn: Brummer; 1801–1803. 2 v. 8° v.1, Om de rödblodige Dyr. Med 7 Kobberplader. [1–10], 429p. 7 pl. v.2, Om de hvidblodige Dyr. Med 7 Kobberplader. [1–10], 512p. 7 pl.

NOTE. N. Möller ög Son, pr. Translation of rec.772, not compared.

774. Cüvier's Elementarischer Entwurf der Naturgeschichte der Thiere aus dem Französischen übersetzt und mit Anmerkungen versehen von C. R. W. Wiedemann, . . . Zwei Bände. Mit 14 Kupfern. Berlin: Joachim Pauli; 1800. 2 v. 14 pl. 8° v.1, ii, xviii, 606p. v.2, 599p.

NOTE. Printer not indicated. Translation of rec.772. Smithsonian Institution Libraries has the only copy located in the United States.

775. Quadro elementar da historia natural dos animaes. Por Mr. Cuvier. Traduzido em portuguez e oferecido A.S.A.R.O Principe R.N.S. Por Antonio d'Almeida, . . . Londres: H. Bryer; 1815. 2 v. 8° v.1, [1815]. xix, 458p. 7 pl. tab.; v.2, 1815. ix, 424p. plus 29 unnumbered pages of index, 7 pl., tab.

NOTE. H. Bryer, pr. Translation of rec.772. The Académie des sciences reported receiving its copy of this at the meeting of November 18, 1816.

776. Lecciones elementales de Historia Natural de los animales, traducidas por D. José Garriga y Baucis. Valencia: Cabrerizo; 1834. 2 v. 8°.

NOTE. Translation of the *Tableau Élémentaire,* rec.772. Not seen. The only copy located is in the Biblioteca nacional, Madrid.

777. Belke, Gustaw. Mastologia czyli historya naturalna zwierząt ssących, obejmująca opisy ich rodzajów, gatunków i odmian; z dodatkiem nauki wyrazów zoograficznych i wiadomości o szczątkach zwierzęcych kopalnych, z dzieł PP. Cuvier, Geoffroy Saint-Hilaire, Pallasa, Illigera, Desmarest, Temminck'a, Lessona, oraz innych znakomitych tegoczesnych Naturalistów zebrana i podług najnowszej metody ułożona przez Gustawa Belke. . . . [Mammalogy, or the natural history of breast feeding animals, including descriptions of genera, species and varieties; and including a glossary of zoological terms and information on animal remains. From the works of Cuvier, Geoffroy St. Hilaire, Pallas, Illiger and other renowned naturalists, collected and composed according to the latest method by Gustaw Belke]. Wilno: Drukiem Jozefa Zawadzkiego; 1847–1850. 4 v. pl. 21cm. v.1,1847. xvii, 383p. 5 pl. v.2,1848. viii, 575p. 5 pl. v.3,1850. 514p. 5 pl. tab.

NOTE. Not seen. The only copy of this located outside of Poland is in the British Library, London. In Poland there is a copy in the Biblioteka Narodowa, Warsaw. Basis for this is apparently the *Règne animal.*

778. Biographie universelle ancienne et moderne, ou Histoire, par ordre alphabétique; de la vie publique et privée de tous les hommes qui se sont distinguées par leurs écrits, leurs actions, leurs talens, leurs crimes. . . Michaud ed., Paris: Michaud Frères; 1811–1862. 5 v. and Suppl.

NOTE. Cuvier was a contributor, but individual articles from the above are not listed in this bibliography. See Dorinda Outram's "The language of natural power: the 'Éloges' of Georges Cuvier and the public language of nineteenth century science," *History of science* 16: 153–178, 1978, for a bibliography of Cuvier's biographical publications.

779. Bowdich, Thomas Edward, 1792–1824. An analysis of the natural classifications of mammalia, for the use of students and travellers. By T. Edward Bowdich, . . . Paris: Printed by J. Smith; 1821. 115p. 15 pl. 8°.

NOTE. "The first part of the text is a translation of almost all but the specific descriptions of the *Mammifères* of Cuvier's *Règne animal*, rendered more perspicuous by a tabular form, interwoven with additions from his Comparative Anatomy, Fossil Remains, the works of Frédéric Cuvier, Duméril, etc., and . . . simplified by explanatory notes and remarks. The second part contains a general outline of the system of Illiger . . . "; cf. Preface, p.iv. The edition of the *Règne animal* referred to is the first, 1817, rec.747.

780. Bowdich, Thomas Edward, 1791–1824. An introduction to the ornithology of Cuvier, for the use of students and travellers. By T. Edward Bowdich, . . . Paris: sold by Treuttel & Würtz; London: Printed by J. Smith; 1821. 86p. 21 pl. 8°.

NOTE. "The text is translated from Cuvier, arranged in a tabular form, illustrated by figures, and explained by notes comprehending the additional information which I have acquired . . . "; cf. Preface. Above is a partial translation from the first edition of the *Règne animal,* 1817, rec.747, but was not compared.

781. Buffon, Georges Louis Leclerc, comte de, 1707–1788. Oeuvres complètes de Buffon, suivies de ses continuateurs Daubenton, Lacépède, Cuvier, Duméril, Poiret, Lesson et Geoffroy-St.-Hilaire. [Par] Buffon et Daubenton. Seule édition complète, avec figures coloriées. Bruxelles: Th. Lejeune; 1828–1833. 14 v. and atlas of 6 v. ports. 742 leaves of pl. v.6 471p. 8°.

NOTE. P. M. de Vroom, pr. No more published, cf. National Union Catalog Pre-1956 Imprints. Cuvier contribution is not identifiable, and it seems probable that most of the additions were not published.

782. Buffon, Georges Louis Leclerc, comte de, 1707–1788. Oeuvres complètes de Buffon, augmentées d'articles supplémentaires . . . par M. le B$^{on.}$ Cuvier. Paris: P. Dupont; 1830–1834. 55 v. and atlas of plates 18°.

NOTE. Not seen. Only location found is for an incomplete set at the Bibliothèque Nationale, Paris.

783. Buffon, Georges Louis Leclerc, comte de, 1707–1788. Oeuvres de Buffon avec les suppléments de Lacépède, Cuvier, Réaumur enrichies d'histoires et d'anecdotes empruntées aux voyageurs français et anglais par M. [A. B.] Morin, . . . Paris: J. Vermot; [1860]. 4 v. 18cm. v.[1], Animaux domestiques. 288p. illus.; v.[2], Carnassiers et rongeurs. 288p. illus.

NOTE. Simon Baçon, pr. Cuvier contributions appear in v.[1], p.221–222 on the hedgehog; and with Lacépède, p.252 on the agouti; in v.[2], p.33 on the jaguar; and with Lacépède, p.176 on the rhinoceros, p.188 on the dromedary, and p.219–220 on the bubal. The catalogue of the Bibliothèque Nationale lists a printing of 2 volumes 12°, 1860, and one with 4 volumes, 1873, 18° (Bibliothèque de la famille). The Catalogue of the British Museum also lists the 2 volume printing, 1860. Neither of these two printings was seen. Each volume of this 4 volume set has 288p. The only copy of the above located in this country is in the Oberlin College Library, Oberlin, Ohio, which has the entire 4 volume set.

784. Buffon, Georges Louis Leclerc, comte de, 1707–1788. Petit Buffon illustré: histoire naturelle . . . extraite des grands ouvrages de Buffon, Lacépède, et Cuvier, par le Bibliophile Jacob [Paul Lacroix]. Paris: Didier; 1861. 4 v. illus. 14cm. [1]. Histoire naturelle des quadrupèdes. 226p. [2]. Histoire naturelle des oiseaux. 257p. [3]. Histoire naturelle des reptiles et des poissons. 291p. [4]. Histoire naturelle des insectes. 268p.

NOTE. P. A. Boureier, pr. Cuvier text not identifiable.

785. Oeuvres de Cuvier et Lacépède, contenant le complément de Buffon à l'histoire des mammifères et des oiseaux l'histoire des cétacés, batraciens, serpents et poissons. Illustrés de 50 planches, environ de 125 sujets coloriés avec le plus grand soin d'après le *Règne animal* de Cuvier, Édition V. Masson Supplément aux Oeuvres complètes de Buffon Annotées par M. Flourens . . . Paris: Garnier Frères; [1883]: v.1, n.d. 668p. 9 pl. 26.5cm. 4 v.

NOTE. A. Quantin, pr. Cuvier text appears in v.1, pp.1–464 and covers birds and mammals with exception of cetacea. Plates for the most part are the same as those in the *Règne animal*, rec.750. Cuvier on p.1 says that this supplement to Buffon is far from presenting all the advances that had been made in the study of quadrupeds since 1789. An earlier publication of the supplement has not been seen, but not all editions of Buffon's *Histoire naturelle*... have been seen.

786. Buffon, Georges Louis Leclerc, comte de, 1707–1788. A natural history of the globe, of man, of beasts, birds, fishes, reptiles, insects and plants. From the writings of Buffon, Cuvier, Lacepede, and other eminent naturalists. Ed. by John Wright . . . A new ed., with improvements from Geoffroy, Griffith, Richardson, Lewis and Clark, Long, Wilson and others . . . Boston: Gray and Bowen, and Philadelphia: Thomas Desilver, Jr.; 1831. 5 v. illus. 16.5cm.

NOTE. Cuvier text, as well as that of other authors cited, is not identifiable. There were a number of other printings of this work in the United States with the same or slightly different titles, not all of which were seen and not all of which indicate Wright on title page. One other printing in England also edited by John Wright, but not seen, is assumed to have unidentifiable contributions.

787. Buffon, Georges Louis Leclerc, comte de, 1707–1788. Natural history of man, the globe, and of quadrupeds, with additions from Cuvier, Lacépède and other eminent naturalists . . . New York: Hurst & Co.; 1879. 2 v. in 1. front., illus., plates.

NOTE. Cuvier text is not identifiable. There are a number of other editons not seen.

788. Choris, Ludovik, 1795–1828. Voyage pittoresque autour du monde, . . . Accompagné de Descriptions de Mammifères et Oiseaux par M. le Baron Cuvier, et M. A. Chamisso, et d'Observations sur les crânes humains, par. M. le docteur Gall. Par M. Louis Choris, peintre. Paris: Imprimerie de Firmin Didot; 1822. vi, [147]p. front. 104 pl., maps. 41.5 × 25.5cm.

NOTE. Text by J. B. B. Eyries. Only one article is attributed to Cuvier, *L'ours gris de l'Amérique septentrionale. Chapeau de bois sur lequel sont peints divers animaux marins,* Pl.V, p.21–22. The latter appears with that title in a number of catalogues under Cuvier as author, but the only indication of Cuvier is as describer of the species. The above work was issued also with publication date of 1820.

789. D., J. Herpétologie ou histoire naturelle des reptiles, d'après G. Cuvier Par M. J. D. et revue Par M. Duméril, . . . Paris: Rue et Place Saint-André-des-Arts, 30; 1834. 100p. 13.5cm. (Bibliothèque populaire).

NOTE. P. Baudouin, pr. Based on second edition of Cuvier's *Règne animal,* rec.748, with direct but unacknowledged quotations. Same text as rec.790.

790. D., J. Herpétologie, ou Histoire naturelle des Reptiles, d'après G. Cuvier, par M. J. D., et revue par M. Duméril . . . Deuxième édition. Paris: Rue et Place Saint-André-des-arts, no. 30; 1834. 104p. illus. 13.5cm. (Bibliothèque populaire; v. 83).

NOTE. Madame Poussin, pr. Text same as that of first edition, rec.789. It is based on the second edition of the *Règne animal,* rec.748. No copy was found in the United States.

791. Demézil, C. Histoire naturelle des poissons, d'après le Baron Ges. Cuvier, par Demézil. Paris: Rue et place St.-André-des-Arts n°30; 1834. 114p. 13.5cm. (Bibliothèque populaire).

NOTE. J.-L. Joly, pr. Published also in 1835, rec.792. Based on the second edition of Cuvier's *Règne animal,* rec.748, with some sections quoted exactly, but not marked as such. The Library of the Academy of Natural Sciences, Philadelphia, has the only copy of this printing located in the U.S.

792. Demézil, C. Histoire naturelle des poissons, d'après Cuvier, par le Dr. Demézil. Paris: Rue et place Saint-André-des-Arts, no.30; 1835. 104p. fig. 14.7cm. (Bibliothèque populaire, no. 99).

NOTE. Madame Poussin, pr. Published also in 1834, rec.791. Based on the second edition of Cuvier's *Règne animal,* rec.748, with some sections directly quoted without acknowledgment. No copy was located in the U.S.

793. Demézil, C. Ornithologie, ou Histoire naturelle des oiseaux, d'après le *Règne animal* de M. le Baron Georges Cuvier, par le docteur Demézil. Paris: Rue et place St.-André-des-Arts nº30; 1832–1833. 2 v. fig. 13.6cm. v.1,1832. 105p., figs.; v.2,1833. 103p., figs. (Bibliothèque populaire).

NOTE. J.-L. Joly, pr. Based on the second edition of the *Règne animal,* rec.748, with some direct but unacknowledged quotes. The Library of the Academy of Natural Sciences, Philadelphia, has the only copy of this printing located in the United States.

794. Demézil, C. Ornithologie ou histoire naturelle des oiseaux d'après le *Règne animal* de Cuvier, par le Dr. Demézil. Deuxième édition. Paris: Rue et place St.-André-des-arts, n.30; 1835. 207p. in 2 pts. illus. 12° Pt.1,1835.; Pt.2, n.d. (Bibliothèque populaire; v. 29).

NOTE. A. Barbier, pr., Pt.1; P. Baudouin, pr., Pt.2. No comparisons were made with other editions. The only copy located was that in the Bayerische Staatsbibliothek, Munich. No copy was seen elsewhere in Europe and none was located in the U.S.

795. Dictionnaire des sciences médicales par une Société de médecins et de chirurgiens: MM. Adelon, Alard, Alibert, Barbier, Bayle, Biett, Bouvenot, Boyer, Breschet, Cadet de Gaissicourt, Cayol, Chaumeton, Chaussier, . . ., Cuvier, . . . Paris: C. L. F. Panckoucke; 1812–1822; v.60.

NOTE. Chaignieau aîné, pr. The title as given in this record does not give all the authors listed on the title page of the book, some 63 in all. Individual articles by Cuvier in the above are included in the Encyclopedia section of this bibliography.

796. Dictionnaire des sciences naturelles, Dans lequel on traite méthodiquement des différens êtres de la nature, considérés, soit en eux-mêmes, d'après l'état actuel de nos connoissances, soit relativement à l'utilité; Suivi d'une biographie des plus célèbres naturalistes: Ouvrage

destiné aux médecins, aux agriculteurs, aux manufacturiers, aux commerçans, et à tous ceux qui ont intérèt à connoître les productions de la nature. Paris: Levrault, Schoell; 1804–1806. 6 v. 8° v.1,1804. lxxij, 560p.; v.2,1804. 515p.; v.3,1804. 492p.; v.4,1805. 483p.; v.5,1806. 480p.; v.6, 1806 atlas of 21? pl. 4°.

NOTE. No printer, v.1; F. G. Levrault, Strasbourg, pr., v.2–5. Only a few copies of v.6 were issued, cf. British Museum (Natural History) Library. *Catalogue* . . . London, v.1, p.458. V.6 was not seen. V.1–5 of above cover 'A' to 'Byt'. Cuvier wrote the Prospectus to the above, a section in the Introduction (rec.824), and, with his collaborator and brother, Frédéric Cuvier, was responsible for "les Articles généraux de l'Histoire naturelle spécialement de la Zoologie; l'Anatomie, la Physiologie, l'Histoire des Reptiles et des Vers, etc." These 5 vols. were reissued, apparently from the same plates, each with an updated addendum, with the balance of the Dictionnaire in 55 vols., in 1816–1830 (rec.797). The Prospectus in v.1 of the above contains a paragraph on style omitted from the later printing (rec.797), and some sections of the introduction also vary in the 2 publications. Atlas of the above was not seen and no complete atlas was located. The Library of the Academy of Natural Sciences, Philadelphia, has the 5 volumes of the above and an atlas of 21 plates. Theirs was the only set of this edition located in the United States. Articles of substance and size by Cuvier are listed in the Encyclopedia section of this bibliography.

797. Dictionnaire des sciences naturelles. Dans lequel on traite méthodiquement des différens êtres de la nature, considérés, soit en eux-mêmes, d'après l'état actuel de nos connoissances, soit relativement à l'utilité. Strasbourg: F. G. Levrault; Paris: Le Normant; 1816–1830; 60 v. 20cm. and atlas 12 v.

NOTE. Cuvier wrote the Prospectus, p.v–xvij of v.1, and also contributed articles on 'Zoologie générale' and 'Anatomie et Physiologie'. "M. F[rédéric]. Cuvier est chargé de la Direction de l'ouvrage . . . "; cf. verso of title page. The Prospectus is for the most part the same as in rec.796 with some changes, including the omission of a paragraph on style. The first five volumes except for preliminary pages in Roman numerals are a reissue apparently from the same plates of rec.796 with an added supplementary update section for each volume. Text of introduction varies. Articles of substance and size contributed by Cuvier are included in the Encyclopedia section of this bibliography. A plan to revise and to update entire set in 1840 did not get farther than v.1, rec.798.

798. ———, Supplément. v.1. (A-Aye). Paris, Strasbourg: F. G. Levrault; 1840–1841. viii, 527p. 20 pl.

NOTE. No more published. Not seen.

799. Drapiez, Pierre Auguste Joseph. Cours complet d'histoire naturelle médicale et pharmaceutique, ou résumé des divers ouvrages concernant l'origine, l'histoire, la description, et les propriétés et l'usage des substances médicamenteuses tirées des trois règnes, publiés par MM. G. Cuvier, A. Richard, Drapiez, etc. Ouvrage dirigé et mis en ordre par ce dernier collaborateur. Bruxelles: H. Dumont, et Londres: Dulau; 1835. 2 v. atlas of 160 pl. 8° v.1, viii, 342p.; v.2, 498p.

NOTE. J. Stienon, pr. Preface states that this includes a résumé of Cuvier's *Règne animal* adapted to the purposes of the editors. This résumé is apparently restricted to the first 77 pages of v.1. V.1 also appears with imprint, Aix-La-Chapelle, Chez L. Kohnen, Librairie, Cologne, même maison, 1835, with Stienon as printer. The only copy of the latter seen was in the Institut de l'histoire naturelle, Bruxelles.

800. [Dujarric de la Rivière, René, ed.]. Cuvier, Sa vie, son oeuvre, Pages choisies présentées par R. Dujarric de la Rivière, . . . Paris: J. Peyronnet; [1969]. 116p. pl. 22cm. (Les maîtres de la biologie).

NOTE. Pr. Imprimerie administrative centrale. Cuvier is generally and correctly listed as the author of this collection, but to assist finding the publication it is entered in this bibliography under

Dujarric. It consists of short excerpts from major publications of Cuvier and brief extracts from some letters. Two letters, possibly published for first and only time, have been separately entered in this bibliography, rec.877 and rec.882.

801. Férussac, André Étienne Just Paschal Joseph François d'Audebard, baron de, 1786–1836. Histoire naturelle générale et particulière des mollusques tant des espèces que l'on trouve aujourd'hui vivantes, que des dépouilles fossiles de celles qui n'existent plus; classés d'après les caractères essentiels . . . Publiée par monographies; Par M. le Baron de Férussac . . . et par plusieurs autres naturalistes. Paris: Arthus Bertrand [et] J.-B. Bailliere; 1834. ix, 95p. 4°.

NOTE. Bourgogne et Martinet, pr. Only 11 livraisons published, cf. British Museum Catalogue of Printed Books. The British Museum Catalogue of Printed Books credits Cuvier with contributions to this. The compiler was unable to identify contributions other than passages enclosed in quotes and attributed to Cuvier with citation of source.

802. Guérin-Méneville, Félix Edouard, 1799–1874. Iconographie du Règne animal de G. Cuvier, ou représentation d'après nature de l'une des espèces les plus remarquables et souvent non encore figurées, de chaque genre d'animaux. Avec un texte descriptif mis au courant de la science. Ouvrage pouvant servir d'atlas à tous les traités de zoologie. Par M. F. E. Guérin-Méneville, . . . Paris et Londres: J. B. Baillière; 1829–1844. 3 v. 2 port., 448 pl. 25cm. v.1. Planches des Animaux vertébrés. 222 pl.; v.2. Planches des Animaux invertébrés. 226 pl.; v.3. Texte explicatif. var. pag. totaling 904p.

NOTE. Fain et Thunot, pr. Based on Cuvier's classification as published in the 2nd edition of the *Règne animal,* rec.748, which is the only Cuvier contribution in this work. Some of the plates were copied in *The animal kingdom,* rec.756. Plates of mollusks were republished in rec.803; of birds in rec.804; and of 'vers' and 'zoophytes' in rec.805. Explanations of plates in these republications are not the same as in the original. Rec.536 is Cuvier's review of this work for the Académie des sciences.

803. [Guérin-Méneville, Félix Edouard, 1799–1874]. Les mollusques décrits et figurés d'après la classification de Georges Cuvier mise au courant des progrès de la science. XXXVI planches représentant en 520 figures Dessinées d'après nature et gravées sur cuivre Les espèces les plus remarquables de ces animaux avec un texte descriptif. Paris: J. B. Baillière et fils; 1868. 74p. 36 pl. 23.5 cm.

NOTE. Corbeil, pr. "Refonte partielle de l'Iconographie," rec.802, cf. Catalogue of the Bibliothèque Nationale, which also lists this under G.-P. Deshayes. With exception of pl.2, plates are same as those in the *Iconographie,* although above lacks last plates of rec.802 in the copy seen. The explanation of plates is a different text. Based on Cuvier's classification from the *Règne animal,* rec.748. In some catalogues it appears under Cuvier as author.

804. [Guérin-Méneville, Félix Edouard, 1799–1874]. Les oiseaux décrits et figurés d'après la classification de Georges Cuvier mise en courant des progrès de la science LXXI planches représentant en 464 figures Dessinées d'après nature et gravées sur cuivre Les espèces les plus remarquables et les caractères génériques tirés du bec et des pattes avec un texte explicatif. Paris: J. B. Baillière et fils; 1869. unnumb.p. 72 pl. 22.5cm.

NOTE. Corbeil, pr. "Refonte partielle de l'Iconographie," rec.802, cf. Catalogue of the Bibliothèque Nationale, which also credits above to Z. Gerbe. Based on classification of Cuvier as presented in the *Règne animal,* rec.748. Appears under Cuvier as author in a number of catalogues. Authorship has also been attributed to Z. Gerbe.

805. [Guérin-Méneville, Félix Edouard, 1799–1874]. Les vers et les zoophytes décrits et figurés d'après la classification de Georges Cuvier mise au courant des progrès de la science XXXVII planches représentant en 550 figures Dessinées d'après nature et gravées sur cuivre Les es-

pèces les plus remarquables de ces animaux avec un texte descriptif. Paris: J. B. Baillière et fils; 1869. 83p. 37 pl. 23.5cm.

NOTE. Corbeil, pr. "Refonte partielle de l'Iconographie," rec.802, cf. Catalogue of the Bibliothèque Nationale. The above is based on classification of Cuvier as presented in the *Règne animal,* rec.748. It appears under Cuvier as author in some catalogues. The Bibliothèque Nationale also credits Paul Gervais with the authorship.

806. Lacépède, Bernard Germain Étienne de la Ville sur Illon, comte de, 1756–1825. La Ménagerie du Muséum National d'Histoire Naturelle, ou les animaux vivants, Peints d'après nature, sur vélin, par le citoyen Maréchal, . . . gravés au Jardin des plantes, avec l'agrément de l'administration, par le citoyen Miger, . . . Avec une note descriptive et historique pour chaque animal, par les citoyens Lacépède et Cuvier, . . . Paris: Miger, Patris, Gilbert, Grandcher & Dentu; An X.—1801–[1805]. [198]p. 41 pl. fol.

NOTE. C. F. Patris, pr. The first edition consisting only of 1 livraison appears under Cuvier in this bibliography, rec.725. Later editions are listed under Lacépède. Above is the 2nd ed. 1st issue, cf. Pieters, below. Also issued [2nd ed. 2nd issue] in 1801–[1805], with the title "La Ménagerie du Muséum National d'Histoire Naturelle, ou description et histoire des animaux qui y vivent ou qui y ont vécu; par les citoyens Lacépède et Cuvier, avec des figures peintes d'après nature, par le citoyen Maréchal, . . . Gravés, . . . par le citoyen Miger, . . ." Of the 37 articles, 31 are by Cuvier, 3 by Étienne Geoffroy, and 2 by Lacépède. For more complete bibliographical description and dating, see Pieters, F. F. J. M. in: *Bijdragen tot de Dierkunde* 51: 219–249, 1981. Numbers of pages and plates are taken from Pieters. Number of pages varied in copies seen by compiler. See rec.725 for description of first edition.

807. Lacépède, Bernard Germain Étienne de la Ville sur Illon, comte de, 1756–1825. La Ménagerie du Muséum National d'Histoire Naturelle, ou description et histoire des animaux qui y vivent et qui y ont

vécu; par les CC. Lacépède, Cuvier et Geoffroy, Avec des Figures peintes, d'après Nature, par Maréchal, . . . , Gravées, avec l'agrément de l'Administration, par Miger, . . . Paris: Miger, Graveur Et Ant. Aug. Renouard, Librairie; XII–1804. 2 v. 16cm. v.1, 353p. 20 pl. v.2, 223p. 21 pl.

NOTE. An octavo edition of rec.806. Title of v.2 varies slightly and publisher is C. F. Patris, Paris. The incomplete first edition is credited entirely to Cuvier, rec.725.

808. Lacépède, Bernard Germain Étienne de la Ville sur Illon, comte de, 1756–1825. La Ménagerie du Muséum d'Histoire Naturelle, ou les animaux vivans, avec des observations curieuses faites sur les individus de chaque espèce; par MM. Lacépède, Cuvier, Geoffroy, et autres savans naturalistes; Ornée de 58 figures, dessinées d'après nature, par M. Maréchal, . . . Gravées par M. Miger, . . . ; Édition originale, augmentée en 1817 de la Description de huit Animaux et de dix-huit gravures, par les mêmes Auteurs. Paris: Tardieu Denesle; 1817. 2 v. pl. 12° v.1. 353p. 27 pl.; v.2. 294p. 31 pl.

NOTE. Imprimerie de Ve. H. Perroneau [n.p.]. For further information, see Pieters in Note to rec.725. For other editions and translations, see rec.725, rec.806, rec.807, rec.809, and rec.810.

809. Lacépède, Bernard Germain Étienne de la Ville sur Illon, comte de, 1756–1825. De Ménagerie van Frankrijk's Nationaal Museum van Natuurlyke Historie, of de leevende dieren, naar de natuur geschilderd op pergament, door den burger Maréchal, . . . , en in het koper gebragt naar de plaaten, door den burger Miger, . . . met goedkeuring van het bestuur, in den Kruidtuin, gegraveerd. Met een beschrijvend en geschiedkundig verslag van ieder dier, door een' natuurkenner. Amsterdam: W. Holtrop; 1801–[1803]. [48]p. 8 pl. fol.

NOTE. Not seen. The Bibliothèque centrale of the University of Amsterdam has the only copy located, discovered by F. F. J. M. Pieters (*Bijdragen tot de Dierkunde* 51: 219–49, 1981, p.245). It is a translation of the first 2 livraisons of the second French edition, rec.806.

810. Lacépède, Bernard Germain Étienne de la Ville sur Illon, comte de, 1756–1825. The wonders of animated nature; consisting of descriptions at large and engraved representations of the principal animals and birds in the Royal Menageries of London and Paris. Translated from La Cepede, with considerable additions by the English editor. London: Richard Phillips; 1810. 198p. 14 pl. 18cm.

NOTE. W. Flint, pr. Translation of rec.806, with omissions and additions. Only two copies have been located, one in the British Museum (Natural History) Library and the other in the Rare Book Room of the Blacker-Wood Library of Zoology and Ornithology, McGill University. The latter is in fragile condition and is available only for limited use. It is the only copy located in North America.

811. Seba, Albert, 1665–1736. Planches de Seba, Locupletissimi Rerum naturalium Thesauri accurata Descriptio. Accompagnées d'un texte explicatif mis au courant de la Science et rédigé par une réunion de Savans: M. Le B$^{on.}$ Cuvier [et al.]; Ouvrage publié Sous les auspices de M. les Professeurs Administrateurs du Muséum royal d'Histoire naturelles de Paris, par les soins de M. E. Guerin . . . Paris et Strasbourg: F. G. Levrault; 1827–1828. 4 v. pl. 46cm.

NOTE. Apparently only v.3 has text, and it consists of an explanation of plates and Cuvier's contribution was not identified in that volume. ". . . le texte n'a été publié que pour les 48 premières planches du tome 3"; cf. Catalogue, V. Durand, Nov. 1841.

ANNOTATIONS

812. [Asiatic Society of Bengal, Calcutta]. Recherches asiatiques, ou mémoires de la Société établie au Bengale Pour faire des recherches sur l'histoire et les antiquités, les arts, les sciences et la littérature de l'Asie; Traduits de l'Anglois par A. Labaume; Revus et augmentés de notes, . . . Et pour la partie des Sciences exactes et naturelles, par MM. Cuvier, Delambre, Lamarck et Olivier, Membres de l'Institut, &. Paris: Imprimerie impériale; An xiv, 1805. 2 v. 27cm. (v.2, 460, 84p.).

NOTE. J. J. Marcel, pr. Notes by Cuvier appear in v.2.

813. Azara, Félix de, 1746–1821. Voyages dans l'Amérique méridionale, Par Don Félix de Azara, . . . depuis 1781 jusqu'en 1801; . . . Publiés d'après les manuscrits de l'auteur, . . . enrichis de notes par G. Cuvier, . . . Paris: Dentu, Imprimeur – Libraire; 1809. 4 v. 25cm. and atlas 4° (v.1. lx, 389p.).

NOTE. Cuvier notes appear in the section on quadrupeds in v.1 and are signed C.V. Published also in Spanish a number of times, see rec.814, and into Italian, rec.816. Some of the notes were also translated unacknowledged into German, rec.814 and rec.815.

814. Azara, Félix de, 1746–1821. Reisen in Süd-Amerika . . . in den Jahren 1781 bis 1801 Aus dem Spanischen mit Anmerkungen und Leben des Verfassers herausgegeben von Walkenaer Nach dem Französischen bearbeitet von W. Lindau. . . . Leipzig: J. C. Hinrichs; 1810: v.1 xiv, 144p.

NOTE. Two of Cuvier's notes on the quadrupeds, translated from rec.813, are included in v.1 without acknowledgment, one condensed and one shortened. They represent the only Cuvier annotations found in the above. Different German translations of the notes were also published, rec.815.

815. Azara, Félix de, 1746–1821. Reise nach Süd=Amerika von Don Felix von Azara in den Jahren 1781 bis 1801. Aus dem Spanischen mit Anmerkungen . . . und herausgegeben von Walkenaer. Aus dem französischen übersetzt von Ch. Weyland, . . . Berlin: Vossische Buchhandlung; 1810. xxvi, 479p. fold. map 8°. (Magazin von merkwürdigen neuen Reisebeschreibungen, aus fremden Sprachen übersetzt und mit erläuternden Anmerkungen begleitet; v. 31).

NOTE. Contains translations from rec.813 of Cuvier footnotes which are attributed to 'the editor'. Cuvier is not acknowledged. Translation is not the same as rec.814.

816. Azara, Félix de, 1746–1821. Viaggi nell'America Meridionale, di D. Felice di Azara . . . fatti da lui tra il 1781 e il 1801. Pubblicati . . .

dal sig. C. A. Walckenaer, tradotti dal sig. Prof. Gaetano Barbieri. Milano: Sonzogno; 1817. 2 v. (Raccolta di viaggi dopo quelli di Cook; v. 33–34).

NOTE. Not seen. Contains annotations by Cuvier, presumably translations of those in rec.813.

817. Azara, Félix de, 1746–1821. Viajes por la América del Sur de don Felix de Azara, . . . desde 1789 hasta 1801. En los quales se da una descripción geográfica, política y civil del Paraguay y del Río de la Plata . . . Con Notas de Mr. G. Cuvier. Montevideo; 1846. 2v. (280p.) 22.7cm. (Biblioteca del Comercio del Plata; v. 2).

NOTE. Caption title. A number of later editions in Spanish also have notes of Cuvier as does an earlier French edition, rec.813, from which this was presumably translated. A few of these notes were published unacknowledged in German, rec.814 and rec.815. An Italian translation of notes was not seen, rec.816.

818. Bowdich, Thomas Edward, 1791–1824. Excursions dans les isles de Madère et de Porto-Santo, faites dans l'automne de 1823, pendant son troisième voyage en Afrique; par feu T. E. Bowdich . . . Ouvrage traduit de l'anglais accompagné de notes de M. le Baron Cuvier et de M. le Baron Humboldt. Paris: Levrault; 1826. 447p. 22cm. and atlas of pl. 4°.

NOTE. Cuvier notes appear in text volume.

819. Camper, Petrus, 1772–1789. Observations anatomiques sur la structure intérieure et le squelette de plusieurs espèces de cétacés. Par Pierre Camper, . . . Publiées par son fils, Adrien Gilles-Camper, . . . , Avec des notes par M. G. Cuvier . . . On y ajoute un Atlas composé de 53 Planches, . . . Ouvrage qui peut faire suite aux "Annales et Mémoires du Muséum d'Histoire Naturelle", et aux "Recherches sur les Ossemens fossiles des Quadrupèdes, par M. Cuvier. Paris: Gabriel Dufour; 1820. 218p. 53 pl. 29.5cm.

NOTE. A. Belin, pr.

820. Delille, Jacques. Les trois règnes de la nature, par Jacques Delille; avec des notes par M. Cuvier, . . . , et autres savants. Paris: H. Nicolle; Giguet et Michaud; Imprimerie des Frères Mame; 1808. 2 v. front. 8°.

NOTE. Published also in *Oeuvres de J. Delille*. Nouv. éd. . . . Paris, L. G. Michaud, 1824, volumes X and XI. Other editions and printings were not investigated. Notes by Cuvier are signed 'Cuvier' and 'C.' A few of the notes are extensive, such as one on fire in Chant I, on p.106–112 of v.1 of above edition and p.53–37 of v.10 of the 1824 *Oeuvres* edition.

821. Plinius Secundus, Caius. Caii Plinii Secundi Historiae naturalis libri xxxvii, etc. Paris: Nicolaus Eligius Lemaire; 1827–1828; Pt.3, Pars tertia continens zoologiam, Georgii Cuvier notis et excursibus illustratam, curante Jo. B. Fr. Steph. Ajasson de Grandsagne, v.3, 4. (Bibliotheca classica Latina . . . ; v. 60, 61).

NOTE. Firman Didot, pr. V.3 includes books 7 and 8; v.4 covers books 9–11. Notes by Cuvier are for most part the same as in rec.823, although some long essays appear in one but not the other. In some cases, these notes are taken from major works of Cuvier and the citations are given.

822. Plinius Secundus, Caius. Histoire naturelle de Pline. Traduction nouvelle Par M. Ajasson de Grandsagne annotée Par MM Beudant Brongniart, G. Cuvier, [etc.]. Paris: Panckoucke; 1829–1830; v.6–8. (Bibliothèque latine-française). Collection des classiques Latins avec la traduction en regard publiée par C. L. F. Panckoucke).

NOTE. Latin and French. An apparent reprint of these three volumes was also published under the title of *Zoologie de Pline*, rec.823, although the two have not been compared. The British Museum Catalogue of Printed Books has what appears to be another issue of this with publication date of 1840. The latter was not seen. Other possible editions or issues were not checked.

823. Plinius Secundus, Caius. Zoologie de Pline, traduction nouvelle, par M. Ajasson de Grandsagne, avec des recherches sur la détermination

des espèces dont Pline a parlé, par M. le Bon. G. Cuvier. Paris: Imprimerie de C. L. F. Panckoucke; 1831. 3 v. 21cm.

NOTE. Latin and French edition. Cuvier's notes are for most part same as those in rec.821. Above is apparently a reprint of rec.822. The two have not been compared. No location in the United States was found for this.

CONTRIBUTIONS TO ENCYCLOPEDIAS

824. De la Zoologie en général. Dictionnaire des sciences naturelles, [rec.797]; 1804; 1: xlij–lxiv.

NOTE. Published also in the later and completed *Dictionnaire des sciences naturelles* . . . rec.797, v.1, 1816, p.l–lij. See Notes to rec.796 and rec.797 for information on the 2 printings.

825. Abdomen (Anat.). Dictionnaire des sciences naturelles, [rec.797]; 1816; 1: 4–6.

NOTE. Originally published in 1804 in rec.796. See Notes to rec.796 and rec.797 for information on the 2 printings.

826. Absorption (Anat.). Dictionnaire des sciences naturelles, [rec.797]; 1816; 1: 77–81.

NOTE. Originally published in 1804 in rec.796. See Notes to rec.796 and rec.797 for information on the 2 printings.

827. Accouplement (Physiol.). Dictionnaire des sciences naturelles, [rec.797]; 1816; 1: 115–117.

NOTE. Published originally in 1804 in rec.796. See Notes to rec.796 and rec.797 for information on the 2 printings.

828. Acéphales (Mollusques, Zool.). Dictionnaire des sciences naturelles, [rec.797]; 1816; 1: 131–134.

NOTE. Originally published in 1804 in rec.796. See Notes to rec.796 and rec.797 for information on the 2 printings.

829. Actinie, Actinea (Zool.). Dictionnaire des sciences naturelles, [rec.797]; 1816; 1: 246–250.

NOTE. First published in 1804 in rec.796. See Notes to rec.796 and rec.797 for information on the 2 printings.

830. Addax (Quad.). Dictionnaire des sciences naturelles, [rec.797]; 1816; 1: 254–255.

NOTE. First published in 1804 in rec.796. See Notes to rec.796 and rec.797 for information on the 2 printings.

831. Allantoïde (Anat.). Dictionnaire des sciences naturelles, [rec.797]; 1816; 1: 477.

NOTE. First published in 1804 in rec.796. See Notes to rec.796 and rec.797 for information on the 2 printings.

832. Alligator. Dictionnaire des sciences naturelles, [rec.797]; 1816; 1: 479–480.

NOTE. Originally published in 1804 in rec.796. See Notes to rec.796 and rec.797 for information on the 2 printings.

833. Alvéole (Anat.). Dictionnaire des sciences naturelles, [rec.797]; 1816; 1: 556–557.

NOTE. Originally published in 1804 in rec.796. See Notes to rec.796 and rec.797 for information on the 2 printings.

834. Ame des bêtes. Dictionnaire des sciences naturelles, [rec.797]; 1816; 2: 30–36.

NOTE. Originally published in 1804 in rec.796. See Notes to rec.796 and rec.797 for information on the 2 printings.

835. Amnios (Anat). Dictionnaire des sciences naturelles, [rec.797]; 1816; 2: 54–55.

NOTE. Originally published in 1804 in rec.796. See Notes to rec.796 and rec.797 for information on the 2 printings.

836. Amphibies (Zool.). Dictionnaire des sciences naturelles, [rec.797]; 1816; 2: 63–64.

NOTE. Originally published in 1804 in rec.796. See Notes to rec.796 and rec.797 for information on the 2 printings.

837. Amphinome. Dictionnaire des sciences naturelles, [rec.797]; 1816; 2: 71–73.

NOTE. Originally published in 1804 in rec.796. See Notes to rec.796 and rec.797 for information on the 2 printings.

838. Amphisbaena. Dictionnaire des sciences naturelles, [rec.797]; 1816; 2: 73–76.

NOTE. Originally published in 1804 in rec.796. See Notes to rec.796 and rec.797 for information on the 2 printings.

839. Anatomie (Anat.). Dictionnaire des sciences naturelles, [rec.797]; 1816; 2: 108–109.

NOTE. Originally published in 1804 in rec.796. See Notes to rec.796 and rec.797 for information on the 2 printings.

840. Animal (Zool.). Dictionnaire des sciences naturelles, [rec.797]; 1816; 2: 158–174.

NOTE. Originally published in 1804 in rec.796. See Notes to rec.796 and rec.797 for information on the 2 printings. Published also in *Dictionnaire des sciences médicales,* Paris, C. L. F. Panckoucke, (rec.795), v.2, 1812, p.142–156, with omission of a small paragraph and a sentence.

841. Anoplotherium. Dictionnaire des sciences naturelles, [rec.797]; 1816; 2(Supplement): 67–68.

842. Anthropolites. Dictionnaire des sciences naturelles, [rec.797]; 1816; 2(Supplement): 80–82.

NOTE. Article is not signed, but author refers to "mes Recherches sur les ossemens fossiles, v.3 et 4" and so presumably is Cuvier.

843. Antilope. Dictionnaire des sciences naturelles, [rec.797]; 1816; 2: 223–251.

NOTE. Originally published in 1804 in rec.796. See Notes to rec.796 and rec.797 for information on the 2 printings.

844. Anus (Anat.). Dictionnaire des sciences naturelles, [rec.797]; 1816; 2: 262–264.

NOTE. Originally published in 1804 in rec.796. See Notes to rec.796 and rec.797 for information on the 2 printings.

845. Aorte (Anat.). Dictionnaire des sciences naturelles, [rec.797]; 1816; 2: 265–266.

NOTE. Originally published in 1804 in rec.796. See Notes to rec.796 and rec.797 for information on the 2 printings.

846. Arenicola. Dictionnaire des sciences naturelles, [rec.797]; 1816; 2: 473–476.

NOTE. Originally published in 1804 in rec.796. See Notes to rec.796 and rec.797 for information on the 2 printings.

847. Artère (Phys. anim.). Dictionnaire des sciences naturelles, [rec.797]; 1817; 3: 156–158.

NOTE. Originally published in 1805 in rec.796. See Notes to rec.796 and rec.797 for information on the 2 printings.

848. Articulation (Anat.). Dictionnaire des sciences naturelles, [rec.797]; 1817; 3: 172–174.

NOTE. Originally published in 1805 in rec.796. See Notes to rec.796 and rec.797 for information on the 2 printings.

849. Azygos (Anat.). Dictionnaire des sciences naturelles, [rec.797]; 1817; 3: 380–381.

NOTE. Originally published in 1805 in rec.796. See Notes to rec.796 and rec.797 for information on the 2 printings. Also published in the *Dictionnaire des sciences médicales*, 1812, v.2 p.515.

850. Bassin (Anat.). Dictionnaire des sciences naturelles, [rec.797]; 1817; 4: 380–381.

NOTE. Originally published in 1806 in rec.796. See Notes to rec.796 and rec.797 for information on the 2 printings.

851. Bec (Zool.). Dictionnaire des sciences naturelles, [rec.797]; 1817; 4: 163–168.

NOTE. Published originally in 1806 in rec.796. See Notes to rec.796 and rec.797 for information on the 2 printings.

852. Boeuf (Mamm.). Dictionnaire des sciences naturelles, [rec.797]; 1817; 5: 18–32.

NOTE. Originally published in 1806 in rec.796. See Notes to rec.796 and rec.797 for information on the 2 printings.

853. Bois (Anat. et Zool.). Dictionnaire des sciences naturelles, [rec.797]; 1817; 5: 36–37.

NOTE. Originally published in 1806 in rec.796. See Notes to rec.796 and rec.797 for information on the 2 printings.

854. Branchie (Ichthyol.). Dictionnaire des sciences naturelles, [rec.797]; 1817; 5: 313–316.

NOTE. Published originally in 1806 in rec.796. See Notes to rec.796 and rec.797 for information on the 2 printings.

855. Cartilage (Anat.). Dictionnaire des sciences naturelles, [rec.797]; 1817; 7: 164–165.

856. Caverneux. Dictionnaire des sciences médicales, [rec.795]; 1813; 4: 385–387.

857. Cerveau (Anat.). Dictionnaire des sciences naturelles, [rec.797]; 1817; 8: 15–17.

858. Dent (Anat.). Dictionnaire des sciences médicales, [rec.795]; 1814; 8: 319–328.

859. Hymen. Dictionnaire des sciences médicales, [rec.795]; 1818; 23: 98–100.

860. Mammouth ou Mammont (Mamm.). Dictionnaire des sciences naturelles, [rec.797]; 1823; 28: 476–479.

861. Nature. Dictionnaire des sciences naturelles, [rec.797]; 1825; 34: 261–268.

NOTE. An English translation with indicated omissions is included in Sarah Bowdich Lee's *Memoirs of Baron Cuvier* 1833, rec.726.

PUBLISHED LETTERS

862. Viénot, John. La Napoléon de l'intelligence Georges Cuvier 1769–1832. Paris: Fischbacher; 1932: 247.

NOTE. Excerpts from numerous letters by Cuvier to various recipients, as well as notes etc. of Cuvier.

863. Letter C [to Louis Agassiz, dated August 3, 1829]. In: Coleman, William. A Note on the Early Relationship between Georges Cuvier and Louis Agassiz. *Journal of the history of medicine,* New York; 1963; 18: 62.

NOTE. English translation of above is published in the same article, p.59–60, and also in Elizabeth C. Agassiz's biography of Louis Agassiz, rec.864.

864. [Letter to Louis Agassiz, dated August 3, 1829]. In: Agassiz, Elizabeth Cary. Louis Agassiz, his life and correspondence. Boston: Houghton, Mifflin; 1885; 1: 114–115.

NOTE. Translation of rec.863.

865. Commune of Bec-aux-Cauchois [dated 14 pluviôse an 2 (February 2, 1794)]. *Société d'Émulation de Montbéliard. Mémoires;* 1969; 67: 24–25 [?].

NOTE. Not seen.

866. Briefwechsel über die fossilen Knochen in der Höhle von Sundwich bei Iserlohn zwischen Hrn. Prof. Benzenberg in Düsseldorf, und

Hrn. Prof. Cuvier in Paris. *Magazin für den neuesten Zustand der Naturkunde,* Jena; 1806; 11: 448–460.

NOTE. Cuvier letters (in German) appear on p.451–452 and 457–460, dated 8 March 1805 and 15 April 1805 respectively.

867. [Letter to Étienne Tessière de Boisbertrand, dated January 29, 1830]. In: Martin, Marietta. Le docteur Koreff. Paris: Champion; 1925: 107.

NOTE. Reprinted, Genève, Slatkine Reprints, 1977.

868. [Letter to J. S. Bonnaire-Mansuy, dated November 22, 1824]. In: *Société d'Émulation de Montbéliard. Mémoires;* 1969; 67: 24–25 [?].

NOTE. Not seen.

869. [Letter to Jean Baptiste Marie Bourgery, 1829]. In: Huard, P. and M. Montagné. Georges Cuvier, 1769–1832, et son temps. *L'extrême-orient médical,* Hanoi; 1949; 1(1 Jan./Apr.): 228–231.

NOTE. Cuvier's assessment of Bourgery's multivolume work on human anatomy.

870. [Letter to Adrien Gilles Camper, 1800?]. In: Coleman, William. Abraham Gottlob Werner vu par Alexandre von Humboldt avec des notes de Georges Cuvier. *Sudhoffs Archiv,* Leipzig; 1963; 57: 477–478.

NOTE. The 'notes' of Cuvier appear as footnotes on pp. 469, 470, 472 of the above.

871. [Lettres à Augustin Pyramus de Candolle, 1802–1812)]. In: Candolle, Augustin Pyramus de. Mémoires et souvenirs. Genève and Paris: Joël Cherbuliez; 1862: 541–554.

NOTE. Six letters (1802–1812, 3 undated) of Cuvier interspersed with letters of others.

872. [Letter to Augustin Pyramus de Candolle, dated April 26, 1831]. In: Candolle, A. P. de. Mort de Cuvier. *Bibliothèque universelle des sciences, lettres et arts,* Genève; 1832; 110: 442–448.

NOTE. Not seen.

873. [Letter to Jean François Champollion, 'le jeune', dated December 5, 1824]. In: Champollion, Aimé Louis. Les deux Champollion Leur Vie et leurs Oeuvres leur correspondance archéologique Relative au Dauphiné et à l'Égypte. Étude complète de biographie et de bibliographie 1778–1867. D'après des documents inédits. Grenoble: Xavier Drevet; 1887: 172–173.

874. [Letter to Victor Cousin, dated December 5, 1820]. In: Barthélemy Saint-Hilaire. *M. Victor Cousin sa vie et sa correspondance*. Paris: Hachette; Félix Alcan; 1895; 1: 113–114.

NOTE. At head of letter: Conseil royal de l'Instruction publique.

875. [Two letters to Emmnuel Crétet, comte de Champmol, the first dated January 8, 1808 and the second is undated]. In: Hamy, E. T. La mission de Geoffroy Saint-Hilaire en Espagne et en Portugal (1808) histoire et documents. *Paris. Muséum national d'histoire naturelle. Nouvelles archives;* 1908; ser.IV v.10: 4–5; 23–24.

876. [Letter to his wife, Anne Marie (Coquet de Trazaile) Duvaucel, dated May 19, 1811, from Pont Sainte Maxence]. In: Lee, Sarah (Wallis) Bowdich, "Mrs. R. Lee". Memoirs of Baron Cuvier. New York: J. & J. Harper; 1833: 122–123.

NOTE. In French with English translation. Appears on p.223–225 in London, Longman, Rees, Brown, etc., 1833, edition of Mrs. Lee's book. Also published in French in the French translation of the above work by Mrs. Lee, *Mémoires du baron Georges Cuvier, publiés en anglais par Mistress Lee, et en français par M. Lacordaire sur les documens fournis par sa famille,* Paris, H. Fournier, 1833, p.235–236.

877. À Madame Cuvier [Anne Marie (Coquet de Trazaile) Duvaucel] à Plombières, [1824?]. In: Cuvier, Georges. Cuvier, sa vie, son oeuvre Pages choisies, présentées par R. Dujarric de la Rivière, . . . [rec.800]. Paris: J. Peyronnet; [1969]: 48.

NOTE. Undated letter to his wife, probably written shortly after Sept. 16, 1824 and probably published in above for the first time. A facsimile of the letter in manuscript appears as pl.7, between pages 64 and 65.

878. [Letter to Joseph Dacier, dated March 20, 1824]. In: Dehérain, Henri. À propos du centenaire. Georges Cuvier, membre de l'Académie des Inscriptions et Belles Lettres. *Journal des savants;* 1932: 223.

879. [Letter to Déodat Silvain Gratet de Dolomieu, dated 16 germinal an 9 (April 16, 1801)]. In: Maindron, Ernest. L'Académie des sciences. Paris: Baillière, Félix Alcan; 1888: 240–241.

NOTE. Letter signed by Berthollet, president of the Institut de France, and Cuvier and Delambre, secretaries, expressing the joy of the members of the Institut at Dolomieu's release after two years' imprisonment.

880. Viénot, John, ed. Lettres inédites de Georges Cuvier à Georges Duvernoy. *Revue chrétienne;* 1905; Ser.II, v.1: 42–57.

NOTE. Eleven letters from 1802 to 1817. Letter #1 of December 6, 1802 on p. 8 was also issued separately, Dole, Girardi et Audebert, [1905], 20p., and published with omissions in Duvernoy, Georges Louis, *Notice historique sur les ouvrages de M. le B[on.] Cuvier . . .*, Paris and Strasbourg, F. G. Levrault, 1833, p.59–61. The latter also contains extracts from other letters to Duvernoy, apparently published in full in above.

881. [Letter to Giovanni Fabbroni, dated March 11, 1807] In: Fabbroni, Giovanni. Scritti di pubblica economia del. Cav. Giovanni Fabbroni. Firenzi: Tipografia di Luigi di Niccolai; 1847; 1: XXXI.

882. [Lettre] à Monsieur Geoffroy, professeur de Zoologie au jardin des plantes à Paris, n.d. [1803?]. In: Cuvier, Georges. Cuvier, sa vie, son oeuvre Pages choisies présentées par R. Dujarric de la Rivière, . . . [rec.800]. Paris: J. Peyronnet; [1969]: 47–48.

NOTE. Undated letter about a trip to Lyons and, with Jussieu, visits to the prefect. A photo reproduction of the manuscript, pl. 6, is on 2 plates between p.48 and p.49.

883. Note instructive du C[itoyen] Cuvier, [dated August 3, 1800]. In: Théodoridès, Jean. Une lettre inédite de Georges Cuvier à la Gesellschaft naturforschender Freunde zu Berlin (1800). *Histoire et biologie;* 1969; fascicule 2: 58–59.

884. [Letter to Pierre Simon Girard, n.d. 1828?]. In: Larronde, Nemours. Cuvier et la géographie. *La Géographie;* 1932; 58: 305.

NOTE. Facsimile of the manuscript letter written on occasion of Cuvier's nomination as president of the Société de Géographie.

885. [Letter to John Gordon, dated August 31, 1817]. In: Ellis, Daniel. Memoirs of the life and writings of John Gordon, MD, . . . Edinburgh: Constable; 1823: 86–87.

NOTE. Letter discusses theories of Franz Josef Gall.

886. Lettres sur l'entomologie [to M.? Hartmann]. *Revue entomologique,* Strasbourg, Paris; 1833; 1: 143–160; 193–210. 2 pl.

NOTE. Translation from German of two letters to M. Hartmann, Nov. 18, 1790, Caen, and May 18, 1791, Fiquainville.

887. [Unsigned letter to Achille d'Héricy, n.d. 1795?]. In: Ardouin, Paul. Georges Cuvier, promoteur de l'idée évolutioniste et créateur de la biologie moderne. Paris: Expansion scientifique française; 1970: 60–64.

NOTE. Photographic reproduction of manuscript letter.

888. [Letter to Johann Hermann, dated July 24, 1795]. In: Duvernoy, Georges Louis. Notice historique sur les ouvrages et la vie de M. le B$^{on.}$ Cuvier. . . . Paris and Strasbourg: F. G. Levrault; 1833: 127–128.

889. [Letter to Johann Hermann, 1800]. In: Duvernoy, Georges Louis. Notice historique sur les ouvrages et la vie de M. le B$^{on.}$ Cuvier, . . . Paris and Strasbourg: F. G. Levrault; 1833: 130–131.

NOTE. Also published in Lee, Sarah (Wallis) Bowdich, "Mrs. R. Lee", *Mémoires du baron Georges Cuvier, publiés en anglais par Mistress Lee, et en français par M. Lacordaire sur les documens fournis par sa famille,* Paris, H. Fournier, 1833, p.29–30. Translated into English, rec.890.

890. [Letter to Johann Hermann, 1800]. In: Lee, Sarah (Wallis) Bowdich, "Mrs. R. Lee". Memoirs of Baron Cuvier. New York: J. & J. Harper; 1833: 17.

NOTE. Translation of rec.889. Published also in a London edition of the above: London, Longman, Rees, Brown etc., 1833, p.27; and in Lauder, Sir Thomas Dick, *The miscellany of natural history: feline species:* Edinburgh, Fraser et al., 1834, [v.2] p.17–18.

891. Une note inédite de Cuvier à Humboldt (1798). In: Théodoridès, Jean. Humboldt et Cuvier. *Biologie médicale;* 1961; 1: LXVI–LXXI.

NOTE. Printed letter and photograph of the original ms. Note is dated October 17, 1798.

892. Lettre de M. G. Cuvier, à M. de Lacépède,... [dated 30 messidor an xiii (July 9, 1805)]. In: Cuvier, Georges. Leçons d'anatomie comparée,... [rec.712]. Paris: Baudouin; 1805; 3: ix–xxviii.

NOTE. Also published in rec.711 in v.3: ix–xxviij; in rec.713, v.4: xiij–xxij; and in rec.714, v.2: ix–xiv. The latter two were not compared.

893. Une lettre de Cuvier [to Pierre Laffin, dated June 8, 1821], Eugène Tissot, ed. *Revue savoisienne,* Annecy; 1876: 54–55.

NOTE. Letter to M. Pierre Laffin in Turin concerning Anoplotherium fossils sent to the Muséum national d'histoire naturelle, Paris.

See rec.517 and rec.562 for letters to Claude Lecoz, Archevêque de Besançon.

894. Lettre de G. Cuvier, . . . , À Jean-Claude Mertrud. In: Cuvier, Georges. Leçons d'anatomie comparée, . . . Paris: Baudouin; [1800] (rec.710); 1: i–xxij.

NOTE. Also published in rec.712, v.1: i–xxij; in rec.713, v.1: xiii–xxxiij with addition of an editor's footnote; and in rec.714, v.1:

ix–xv. Translated into German, rec.895. Extracted in rec.800, p.61–64.

895. Brief von G. Cuvier, . . . , an Herrn Mertrud, . . . In: Cuvier, Georges. Vorlesungen über vergleichende Anatomie, . . . Leipzig: Paul Gotthelf Kummer; 1809; 1 (rec.719): v–xxvi.

NOTE. Translation of rec.894.

896. George Cuvier's Briefe an C. H. Pfaff aus den Jahren 1788 bis 1792, naturhistorischen, politischen und literarischen Inhalts. Nebst einer biographischen Notiz über G. Cuvier von C. H. Pfaff. Herausgegeben von Dr. W. F. G. Behn, . . . Kiel: Schwers'sche Buchhandlung; 1845. xvi, 309p. 6 pl. 8°.

NOTE. Letters written by Cuvier in German. Translated into French, rec.897, but not compared with original German. Very brief extracts in French of letters of October 14 and November 17, 1788 were also published, rec.898. See also Note to rec.897 for further extracts in French.

897. Lettres de Georges Cuvier à C. M.[i.e. H.] Pfaff sur l'histoire naturelle, la politique et la littérature 1788–1792 traduites de l'allemand par Louis Marchant . . . Paris: Victor Masson; 1858. 314p. 1 pl. 8°.

NOTE. Corbeil, Crete, pr. Translation of letters written in German by Cuvier, rec.896. Translation was not compared with original German. Extracts from some of the letters have also been published, rec.898 and 899.

898. [Extracts of letters to C. H. Pfaff, dated October 14, 1788 and November 17, 1788]. In: Cuvier, Georges. Cuvier, sa vie, son oeuvre Pages choisies présentées par R. Dujarric de la Rivière, . . . [rec.800]. Paris: J. Peyronnet; [1969]: 18–21.

NOTE. The complete letters were published in rec.897 on p.57–66, and 69–75, respectively. Extracts of the October 14 and November 11, 1788 letters were also published, rec.898, but were not compared with the above.

899. Lettres de Cuvier à Pfaff. In: Ardouin, Paul. Georges Cuvier, promoteur de l'idée évolutioniste . . . Paris: Expansion scientifique française; 1970: 43–45.

NOTE. Extracts from letters to Pfaff on October 14, 1788, November 11, 1788, February 2, 1789 as well as a complete letter with one small omission, dated June 22, 1789. The complete letters were published, rec.897. Extracts of October 14,1788 and November 17, 1788 were also published, rec.898, but not compared with above.

900. De insectis aliquot, Epistola-Zoologico critica, ad DD. C.F.[i.e.H.?] Pfaff & E. d. Marschall. Pfaffio Marschalloque Suis, S. p. d. G. L. Cuvier. *Stettiner entomologische Zeitung,* Stettin; 1844; 5: 6–9.

NOTE. Letter in Latin dated Oct. 16, 1788.

901. [Letter to M. Raguet l'Épine, Député de Loir-et-Cher . . .]. In: Tamizey de Larroque, Phillipe, ed. Une petite gerbe de billets inédits . . . *Bulletin du bibliophile;* 1890: 50–52.

NOTE. Signed "Cuvier" and undated. The editor identifies the letter writer as the author of the *Discours sur les révolutions* . . . in a footnote on p.52. In that footnote and one on p.51, he is critical of Cuvier textual errors. The entire collection of letters edited by Tamizey was also published as a monograph with the same title, *Une petite gerbe de billets inédits* . . . Paris, Techener, 1890, with Cuvier letter on p.15–17.

902. Lettres de Georges et Anne Cuvier à l'abbé Camillo Ranzani (1812–1827). In: Théodoridès, Jean. Quelques documents inédits ou peu connus relatifs à Georges Cuvier, à sa famille et à son salon. *Stendhal Club. Revue trimestrielle;* [1966–1967]: 58–64.

903. [Letter to William Swainson], dated May 6, 1821. In: Swainson, William. Testimonials presented to the trustees of the British Museum on behalf of William Swainson . . . and accompanying his application for the appointment of assistant keeper in the natural history department vacated by the resignation of Dr. Leach in February 1822. London; 1822.

NOTE. Not seen. See Dorinda Outram, *The letters of Georges Cuvier . . .*, 1980 p.89, no. 232/1.

904. Le Secrétaire perpétuel pour les sciences naturelles, à M. Toulongeons, . . . In: Toulongeon, François-Emmanuel. Mémoire sur les usines établies sur les petites rivières. Lu à la Ire. classe de l'Institut, Dans la séance du 11 avril 1807. [Paris: Imprimerie de Dentu; 1807]: [i]–ij.

NOTE. Cuvier's letter appears with heading: Institut National. Classe des sciences physiques et mathématiques. Paris, le 19 mai 1807.

905. [Letter to Achille Valenciennes, n.d. 1827?]. In: Lee, Sarah (Wallis) Bowdich, "Mrs. R. Lee". Mémoires du baron Georges Cuvier, publiés en anglais par Mistress Lee, et en français par M. Théodore Lacordaire, sur les documens fournis par sa famille. Paris: H. Fournier; 1833: 236–237.

NOTE. Translated into English in rec.906. A facsimile of the manuscript was also published as the frontispiece in Lee, Sarah (Wallis) Bowdich, "Mrs. R. Lee", *Memoirs of Baron Cuvier,* London, 1833. There are some differences in wording between the above and the manuscript.

906. [Letter to Achille Valenciennes, n.d. 1827?]. In: Lee, Sarah (Wallis) Bowdich, "Mrs. R. Lee". Memoirs of Baron Cuvier. New York: J. & J. Harper; 1833: 124.

NOTE. English translation of rec.905. Also published in the English edition of above, London, Longman, Rees, Brown, etc., 1833, p.226.

907. [Letter to Charles François Dominique de Villers, dated 18 floréal 10 (May 8, 1802)]. In: Isler, Meyer, ed. Briefe . . . Auswahl aus dem handschriftlichen Nachlasse des Ch. de Villers herausgegeben von M. Isler. Hamburg: Otto Meissner; 1879: 60–61.

908. [Letter to James Wilson]. In: Lauder, Thomas Dick. Miscellany of natural history: feline species. Edinburgh: Fraser, and London: Smith Elder, and Dublin: W. R. Wakeman; 1834; [2]: 41–42.

NOTE. English translation follows above on p. 42.

RANDOM STATISTICAL NOTES

Considering that Cuvier juggled as many balls as he did, his productivity is impressive. He published at least 349 primary articles, from which 455 republications and translations were derived. A count of the éloges listed by Dorinda Outram[1] adds another 80 to the primary total.[2] In addition, Cuvier contributed a total of 38 articles to the *Dictionnaire encyclopédique des sciences naturelles* and the *Dictionnaire des sciences médicales*. He published 10 major monographs, most multivolume; and prepared 4 new monograph editions, as well as writing many shorter monographic works and annotations.

Unless otherwise indicated, the statistics in this section are taken from the Journal and *Analyses des travaux. . .* sections above, which include the articles and ephemeral or pamphlet publications in the bibliography. Monographs have been omitted in this analysis largely because the timeliness and abundance of journal literature better display the trends and shifts in Cuvier's interests and productivity and also, to some extent, the general directions of natural history research of the time. Statistics for republication of primary articles[3]—comprised of same-text

---

[1] Outram, D. "The language of natural power: the 'Éloges' of Georges Cuvier and the public language of nineteenth century science," *History of science,* 1978, 16:153–178.

[2] Statistics for the éloges were not included in any other statistics in this bibliography.

[3] *Primary* is defined as an article that has not been derived from any other article or monograph. The term *article* is understood to include such other works as speeches and lectures.

republications, extracts, summaries, condensations, augmentations, and translations—also tell a story of trends in popularity and of those of Cuvier's contributions that were most appreciated by his contemporaries.

Of the 351 primary articles in the bibliography, 68% were republished. Counting the various forms of republication, 17% were republished only once, while 11% were republished five or more times. In arriving at these statistics, republication of an article in his own monographs, such as *Recherches sur les ossemens fossiles de quadrupèdes* or *Mémoires pour servir à l'anatomie des mollusques,* was not counted. This was republication by Cuvier himself.

The most frequently republished in the journal literature of Cuvier's work were the various publications of his popular lecture series on the history of the natural sciences delivered in 1829–1830, rec.483, republished eight times as articles (and four times as monographs). Three of his papers were republished seven times each: another work on the history of science, rec.318; the second paper of the debate with Geoffroy Saint Hilaire, rec.495; and his critique, rec.471, on François Roulin's memoir on a new species of tapir, a harbinger of the great debate between Cuvier and Geoffroy. A memoir on the eggs of the cuttlefish, published a month before his death in 1832, was republished six times, rec. 537.

Cuvier's scientific reputation spread abroad early in his career. Before 1801, 17 translations of his articles had appeared. Overall, his journal and annual review articles were translated 174 times into eight languages from 112 French originals, 6 of which were excerpts from his monographs. Perhaps because of his education in Germany and his acquaintance with scientists there, Cuvier was translated more frequently into that language than into any other, with 101 translations. English was the second language with 58, followed by 6 translations into Italian, 5 into Polish, and 1 each into Hungarian, Romanian, Russian, and Spanish. Apparently the annual reviews of science were of greater interest in England than in Germany. These, or extracts from them, were translated only 6 times into German, but 21 times into English. Of the 64 translations of monographs or portions of monographs, 30 were into

English, 16 into German, 4 into Italian, 3 into Spanish, 2 each into Polish and Russian, and 1 each into Czech, Danish, Dutch, Hungarian, Portuguese, Swedish, and Welsh.

His most translated paper, rec.60 on fossil quadrupeds, was published in 1800 and translated four times—once each into English and Italian, and twice into German. Translated three times were rec.22 in 1796, his first paper on the *Megatherium,* translated into English, German, and Spanish; rec.44, another paper on fossil quadrupeds, published in 1798, translated twice into English and once into German; and extracts from his public lecture series on the history of the natural sciences, rec.483, also translated twice into English and once into German.

Surprisingly, the only translation found for any of the papers of the famous Geoffroy/Cuvier debate, republished so many times in France, was an English translation of the first paper of the debate. This paper alone was republished seven times in France.

Cuvier's most productive year was 1804, when he published 27 primary papers, all describing his own research. In addition he supplied 17 contributions to the *Dictionnaire des sciences naturelles,* Volume 1, published in the same year.

In 1804 he was thirty-five years old, had been married in February, and was recently back from two trips to southern France, where he had been sent as inspector general of public education to organize lycées. In compliance with an ordinance of the government (Napoleon), he was writing the report on the natural sciences for a review of progress in the sciences and arts since 1789—a task that, he complained, had cost him what amounted to a year's work and also required him to force himself to remain thorough and impartial. His report was one of the earliest submitted (1808), and of the three submitted early, Cuvier considered it to have been the one that "least poorly" fulfilled the objective. During 1804, he was also working with Duvernoy for publication in 1805 of the last three volumes of his *Leçons d'anatomie comparée* (rec.712). In 1804 he carried fewer of the responsibilities of his later years, although he already held a number of teaching positions and was perpetual secretary of the Académie des sciences. It was also the year of the birth and death of his first child.

His leanest years were 1811, 1814, 1816, and 1823. In 1811 he was sent by Napoleon to the Netherlands and southern Germany to study and reorganize public education. In 1816 he was preparing for the publication of his *Règne animal,* received by the Académie des sciences on December 2, 1816. In 1814 and 1823 it may have been the burden of increased activities and responsibilities that distracted his attention from research.

Cuvier was criticized in his lifetime by his scientific peers for spreading himself too thin in his many activities—reducing the amount and even the quality of his research as well as neglecting natural history in general. Considering the number of articles reporting his own research, this criticism would appear to be justified. From 1792 through 1809, 17 years, he published 136 original research papers, 76% of his total primary publication output for this period. From 1810 through 1832, 23 years, he published 41 research papers, 27% of his total papers published during the period. His non-research papers include papers on the history of science, reports and addresses required for the various offices he held, speeches before the Chambers of Peers and Deputies, and, most of all, his critiques and reviews of the works of others.

Even excluding his annual review of science, Cuvier's papers assessing the works of others, whether their written articles or collections made by them, were popular choices for republication. All except eight of these were assessments by commissions appointed by the Académie des sciences to review a work or collection presented and, if merited, to recommend endorsement by the Académie. Cuvier was presumably the author of those reports for which he served as reporter. Of these reports, 85 are primary publications and 105 are republications derived from them. Almost all of those read before the Académie appeared in the manuscript minutes (*Académie des sciences, Procès verbaux*), and, if not already published elsewhere, were first disseminated in the early twentieth century, when the *Procès verbaux* were published for the first time.

Cuvier attracted the attention of newspapers especially after he became a public figure. This interest included all aspects of his career:

scientific research, lectures, speeches and debates before the Chambers of Deputies and of Peers, and publications resulting from the various offices he held. Most popular with newspapers were his critiques of the works of others. Publication of these amounted to 44% of the total material by Cuvier chosen for publication by the newspapers examined. Second in popularity with newspapers were articles on his own scientific research, making up 23% of the total.[4]

Some of his memoirs reporting his own natural history research appeared first in newspapers and were published later in scientific journals. All of those reported by newspapers were read before the Académie, and some were published within a few days verbatim or condensed. Whether Cuvier would give a paper first to the newspapers, which, after publishing it, would pass it on to scientific journals; whether he supplied copies (necessarily handwritten) to both; or whether journals copied from newspapers is not clear. David A. Kronick speaks of the "almost complete freedom with which scholarly journals borrowed from one another, often without indicating their sources."[5]

With Cuvier's reputation, at least among twentieth-century scholars, for masterly self-promotion, it would be of some interest to know if the overall multiplication of publication of his papers was authorized or possibly even promoted.

Impressive as it appears to be, it is not possible to evaluate the uniqueness of his output, nor his influence on the science of his day, as measured by numbers of republications and translations, without having comparable statistics for other contemporary scientists. Certainly his active participation and accomplishments in so many different fields are unusual, even without considering the time-consuming activities involved in the business of living at that time and the degree of limitation on productive time—handwritten papers or copies, hand-produced illus-

---

[4] Five newspapers were searched at the Bibliothèque nationale or on microfilm. Not all issues of those searched were available. The major newspapers searched were *Le Moniteur universel, Le Globe, Le National, Le Temps,* and *L'Universel,* all published in Paris.

[5] "Authorship and authority in the scientific periodicals of the seventeenth and eighteenth centuries," *The library quarterly,* 1978, 48:255–275, p. 270.

trations, travel near or far by horse, and communication in person or by letter. As many of those who knew and worked with him observed, Cuvier adjusted his life to make every minute count. Possibly in the long run, these deficiencies contributed to his productivity by eliminating the habit of wasted time.

PERSONAL NAME INDEX

Entries are indexed by record number. Other signers of reports made to the Académie des sciences assessing the work of others and for which Cuvier was the reporter are not included.

SUBJECT INDEX

Entries are indexed by record number. Subjects are taken from keywords in the title. Geographic subjects are not included unless they are the major subject of the title. Titles of major monographs are included.